Leckie✕Leckie
Scotland's leading educational publishers

Practice Papers for SQA Exams

Standard Grade | General

Biology

Text © 2009 Lorraine Clark and Rebecca Shearer
Design and layout © 2009 Leckie & Leckie

01/150609

ISBN 978-1-84372-789-7

Published by
Leckie & Leckie Ltd, 4 Queen Street, Edinburgh, EH2 1JE
Tel: 0131 220 6831 Fax: 0131 225 9987
enquiries@leckieandleckie.co.uk   www.leckieandleckie.co.uk

A CIP Catalogue record for this book is available from the British Library.

Leckie & Leckie Ltd is a division of Huveaux plc.

Questions and answers in this book do not emanate from SQA. All of our entirely new and original Practice Papers have been written by experienced authors working directly for the publisher.

Introduction

Layout of the Book

This book contains practice exam papers, which mirror the actual SQA exam as much as possible. The layout, paper colour and question level are all similar to the actual exam that you will sit, so that you are familiar with what the exam paper will look like.

The answer section is at the back of the book. Each answer contains a worked out answer or solution so that you can see how the right answer has been arrived at. The answers also include practical tips on how to tackle certain types of questions, details of how marks are awarded and advice on just what the examiners will be looking for.

Revision advice is provided in this introductory section of the book, so please read on!

How To Use This Book

The Practice Papers can be used in two main ways:

1. You can complete an entire practice paper as preparation for the final exam. If you would like to use the book in this way, you can either complete the practice paper under exam style conditions by setting yourself a time for each paper and answering it as well as possible without using any references or notes. Alternatively, you can answer the practice paper questions as a revision exercise, using your notes to produce a model answer. Your teacher may mark these for you.

2. You can use the Topic Index at the back of this book to find all the questions within the book that deal with a specific topic. This allows you to focus specifically on areas that you particularly want to revise or, if you are mid-way through your course, it lets you practise answering exam-style questions for just those topics that you have studied.

Revision Advice

Work out a revision timetable for each week's work in advance – remember to cover all of your subjects and to leave time for homework and breaks. For example:

Day	6pm–6.45pm	7pm–8pm	8.15pm–9pm	9.15pm–10pm
Monday	Homework	Homework	English Revision	Chemistry Revision
Tuesday	Maths Revision	Physics Revision	Homework	Free
Wednesday	Geography Revision	Modern Studies Revision	English Revision	French Revision
Thursday	Homework	Maths Revision	Chemistry Revision	Free
Friday	Geography Revision	French Revision	Free	Free
Saturday	Free	Free	Free	Free
Sunday	Modern Studies Revision	Maths Revision	Modern Studies	Homework

Make sure that you have at least one evening free a week to relax, socialise and re-charge your batteries. It also gives your brain a chance to process the information that you have been feeding it all week.

Arrange your study time into one hour or 30 minutes sessions, with a break between sessions e.g. 6pm–7pm, 7.15pm–7.45pm, 8pm–9pm. Try to start studying as early as possible in the evening when your brain is still alert and be aware that the longer you put off starting, the harder it will be to start!

Study a different subject in each session, except for the day before an exam.

Do something different during your breaks between study sessions – have a cup of tea, or listen to some music. Don't let your 15 minutes expand into 20 or 25 minutes though!

Have your class notes and any textbooks available for your revision to hand as well as plenty of blank paper, a pen, etc. You may like to make keyword sheets like the Biology example below:

Keyword	Meaning
population	number of organisms of one type
photosynthesis	process by which plants make their own food
phenotype	physical appearance of an individual

Finally forget or ignore all or some of the advice in this section if you are happy with your present way of studying. Everyone revises differently, so find a way that works for you!

Transfer Your Knowledge

As well as using your class notes and textbooks to revise, these practice papers will also be a useful revision tool as they will help you to get used to answering exam style questions. You may find as you work through the questions that they refer to an experiment or an example that you haven't come across before. Don't worry! You should be able to transfer your knowledge of a topic or theme to a new example. The enhanced answer section at the back will demonstrate how to read and interpret the question to identify the topic being examined and how to apply your course knowledge in order to answer the question successfully.

Command Words

In the practice papers and in the exam itself, a number of command words will be used in the questions. These command words are used to show you how you should answer a question – some words indicate that you should write more than others. If you familiarise yourself with these command words, it will help you to structure your answers more effectively.

Command Word	Meaning/Explanation
Name, state, identify, list	Giving a list is acceptable here – as a general rule you will get one mark for each point you give
Suggest	Give more than a list – perhaps a reason or an idea
Describe	Give more detail than you would in a list and use values where you can
Explain	Discuss WHY an action has been taken or an outcome reached – what are the reasons and/or processes behind it.
Justify	Give reasons for your answer, stating why you have taken an action or reached a particular conclusion.
Define	Give the meaning of the term.
Compare	Describe both items and ALSO say how they are different or similar to each other

In the Exam

Watch your time and pace yourself carefully. Work out roughly how much time you can spend on each answer and try to stick to this.

Be clear before the exam what the instructions are likely to be
Biology is straight forward. You are expected to answer EVERY question.
DON'T MISS ANY QUESTIONS OUT.

The practice papers will help you to become familiar with the exam's instructions.

Read the question thoroughly before you begin to answer it – make sure you know exactly what the question is asking you to do.

There are often clues in the paragraph/ sentence at the start of a question. Don't forget to read this section.

Don't repeat yourself as you will not get any more marks for saying the same thing twice. This also applies to annotated diagrams which will not get you any extra marks if the information is repeated in the written part of your answer.

Give proper explanations. A common error is to give descriptions rather than explanations. If you are asked to explain something, you should be giving reasons. Check your answer to an 'explain' question.

Look carefully at the number of marks that a question is worth – if it is worth 2 marks then a one word answer will not be sufficient.

If you have to extract information from a table, make sure you are using the correct values in your answer. If there are 2 lines on a graph make sure you are taking information from the right line.

Problem solving is an important aspect and so calculations will have to be done. Here are a few pointers

PERCENTAGES:

To express a number as a percentage of the total

Number / Total × 100

To calculate a % increase

Increase in value / starting value × 100

To calculate a % decrease

Decrease in value / starting value × 100

AVERAGES

Add up all the values and divide by the number of values

RATIOS

Write down the numbers to be used then divide each number by the same number to make each number smaller. The ratio will always be a whole number.

Good luck!

Topic Index

Topic	Paper 1	Paper 2	Paper 3	Paper 4
Biosphere				
(a) Investigating an ecosystem		Q.2	Q.1	Q.1
(b) How it Works	Q.1	Q.1, 3	Q.2, 4	Q.2
(c) Control and Management	Q.2		Q.5	
World of Plants				
(a) Introducing Plants	Q.3	Q.4(b)		
(b) Growing Plants		Q.4(a)	Q.12, 13	Q.3
(c) Making Food	Q.5			Q.4, 5
Animal Survival				
(a) Need for Food	Q.6	Q.5		Q.8
(b) Reproduction			Q.6	Q.10 PS
(c) Water and Waste	Q.7			Q.7
(d) Responding to the Environment			Q.7, 8PS, 9	
Investigating Cells				
(a) Investigating Living Cells	Q.8			Q.6(a)&(b)
(b) Investigating Diffusion	Q.9		Q.1O	Q.6(c)
(c) Investigating Cell Division	Q.12	Q.8		
(d) Investigating Enzymes	Q.10	Q.6		Q.9
(e) Investigating Aerobic Respiration			Q.11	
Body in Action				
(a) Movement	Q.11			Q.11
(b) Need for Energy	Q.13	Q.10	Q.16	Q.12
(c) Coordination		Q.9	Q.14	Q.13
(d) Changing Levels of Performance	Q.14			
Inheritance				
(a) Variation		Q.13		
(b) What is Inheritance?	Q.16	Q.12	Q.17	Q.15, Q.16(a)&(b)
(c) Genetics and Society				Q.16(c)
Biotechnology				
(a) Living Factories	Q.17(a)			Q.17, 21
(b) Problems and Profit with Waste	Q.17(b)	Q.14, 15		
(c) Reprogramming Microbes			Q18, 19	Q.20
Problem Solving	Q.4, 15	Q.7, 11, 16	Q3, 15	Q.14, 18, 19, 21

Exam A

Biology

Standard Grade: General

Practice Papers
For SQA Exams

Exam A
General Level

Fill in these boxes:

Name of centre

Town

Forename(s)

Surname

You have 1 hour 30 minutes to complete this paper.

Try to answer all of the questions in the time allowed.

Write your answers in the spaces provided, including all of your working.

Leckie×Leckie

Scotland's leading educational publishers

	KU	PS

1. (a) The diagram below shows a food web from a field ecosystem.

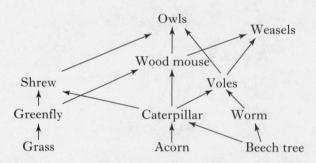

(i) What do the arrows in the food web represent?

direction of energy flow ✓

1

(ii) Name a producer.

Grass ✓

1

(iii) Name a consumer in the food web.

Greenfly ✓

1

(b) (i) Use the food web to complete the food chain below, consisting of four organisms.

Grass ⟹ *Green Fly* ⟹ *Shrew* ⟹ *Owls* ✓

1

(ii) How many food chains in the food web involve voles?

two ✓

1

(iii) Ladybirds feed on greenfly and are eaten by shrews.

Add ladybirds to the food web to show their feeding relationships.

Shrew

Ladybird

Greenfly

1

2. The table shows the numbers of four different organisms at two sites (A and B) in a river.

Site	Mayfly larvae	Shrimps	Water lice	Leeches
A	4	36	86	100
B	80	140	4	10

(a) Which organism has the greatest difference in number between sites A and B?

~~Maythm~~ *Shrimps.*

(b) It was suggested that sewage from an agricultural facility had been discharged into the river at Site A.

Agriculture is one source of pollution.
Name another source.

_____

(c) Mayfly larvae prefer water with high levels of oxygen.

Put a tick beside the line in the table that is correct for the oxygen content of the water at sites A and B.

Oxygen content of water		Tick the correct line
Site A	Site B	
Low	Low	
Low	High	
High	High	
High	Low	

(d) Give the ratio of leeches found at site A to the number of leeches found at site B.

leeches at site A: _____

_____

leeches at site A: _____

_____

KU	PS
	1
1	
	1
	1
	1

	KU	PS

3. Plants can be used in a variety of ways by humans.

    (a) Give two uses of plants by humans.

    _____ **1**

    (b) Give an example of a food made from plants.

    _____ **1**

    (c) Give one raw material that plants can provide.

    _____ **1**

4. Read the following passage and answer the questions based on it.

GROWTH IN PLANTS.

Trees and plants are growing bigger and faster in response to the billions of tons of carbon dioxide released into the atmosphere by humans.

The increased growth has been discovered in a variety of plants, ranging from tropical rainforests to British sugar beet crops.

It means they are soaking up at least some of the carbon dioxide that would otherwise be accelerating the rate of climate change. It also suggests the potential for higher crop yields.

Since 1750 the concentration in the air of carbon dioxide has risen from 278 parts per million (ppm) to 380 ppm, making it easier for plants to acquire the carbon dioxide needed for rapid growth.

The experiments generally suggest that raised carbon dioxide levels, similar to those predicted for the middle of this century, would boost the yields of mainstream crops, such as maize, rice and soy, by about 13%.

However, scientists have warned against drawing false comfort from such findings. They point out that although levels will boost plant growth, other factors are also increasing that are associated with climate change, such as rising temperatures and drought. These are likely to have a negative effect.

KU | PS

(a) Give an example of a plant that has shown increased growth due to carbon dioxide being released into the environment.

_____

1

(b) What was the starting concentration of carbon dioxide in the air in 1750?

_____

1

(c) How much has the concentration of carbon dioxide in the air increased by since 1750?

Space for calculation

_____

_____ ppm

1

(d) Give two examples of mainstream plants.

_____

1

(e) Give a negative effect associated with climate change.

_____

1

(f) Give a possible advantage of increasing levels of carbon dioxide?

_____.

1

5. Plants use photosynthesis to produce glucose.

(a) Complete the word equation for photosynthesis.

Water + ... ⟶ .... + glucose

(b) What is the source of energy for this process?

_____

_____

(c) A student was interested in finding the cheapest way to grow tomatoes in the winter.

He investigated the relationship between light intensity and the rate of photosynthesis.
He took 7 tomato plants that had all grown 5 leaves.
He placed them in a sealed glass case and measured how much carbon dioxide was used by the plant during photosynthesis.

The heat shield lets light but not heat through.

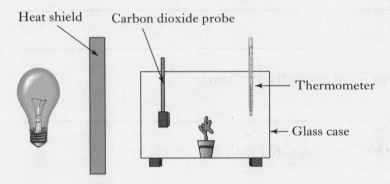

Heat shield   Carbon dioxide probe
Thermometer
Glass case

The results he obtained are in the table below.

Light intensity (% of normal light)	CO_2 concentration at start (units)	CO_2 concentration after 12 hours (units)	CO_2 concentration used by plant (units)
10	35	34.2	0.8
20	35	30.6	4.4
40	35	29.4	6.1
60	35	28.2	6.8
80	35	28.0	7
95	35	27.9	7.1

(i) What is the average CO_2 concentration used by the plant?

_____

_____

KU: 1

KU: 1

PS: 1

(ii) Plot a line graph of the data on the grid below.

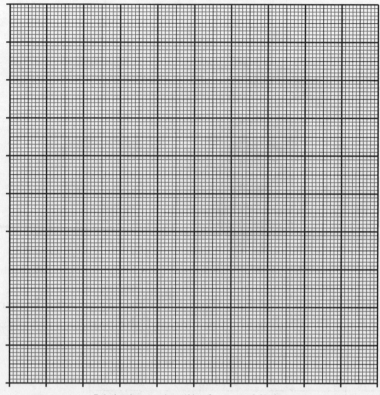

Light intensity (% of normal light)

(d) What is the variable being changed in this experiment?

_____.

(e) Name one variable that should be kept constant in the experiment

_____

(f) In the experiment design, one plant was placed at each light intensity.

Was this a good idea? _____

Reason _____

_____

(g) The student was disappointed with his results.

He said 'I still don't know exactly what light intensity is best to use'.

How could he improve his design to solve this problem?

_____

_____

KU	PS
	2
	1
	1
	1
	1

6. The diagram below shows the human digestive system.

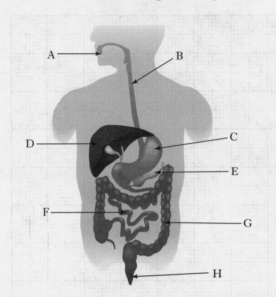

(a) Complete the table by entering the information that is missing by writing, either the letter from the diagram or the name of the part.

Letter	Part of digestive system
	oesophagus
D	
G	
	rectum

(b) **Using a letter** from the diagram, write down where the absorption of digested food into the bloodstream takes place.

_____

(c) Describe two ways that the small intestine is related to its function.

_____

_____

_____

(d) Fibre is an important part of a healthy diet.

The fibre content of a variety of foods is shown in the table below.

Foodstuff	Fibre content (%)
Wholemeal bread	6
Bran flakes	48
Spinach	60
Wheat flakes	12
Carrots	4
Apple	2
Peas	8

KU | PS

2

1

2

(i) On the grid below, complete the axis and the bars to show the fibre content of each of the foodstuffs.

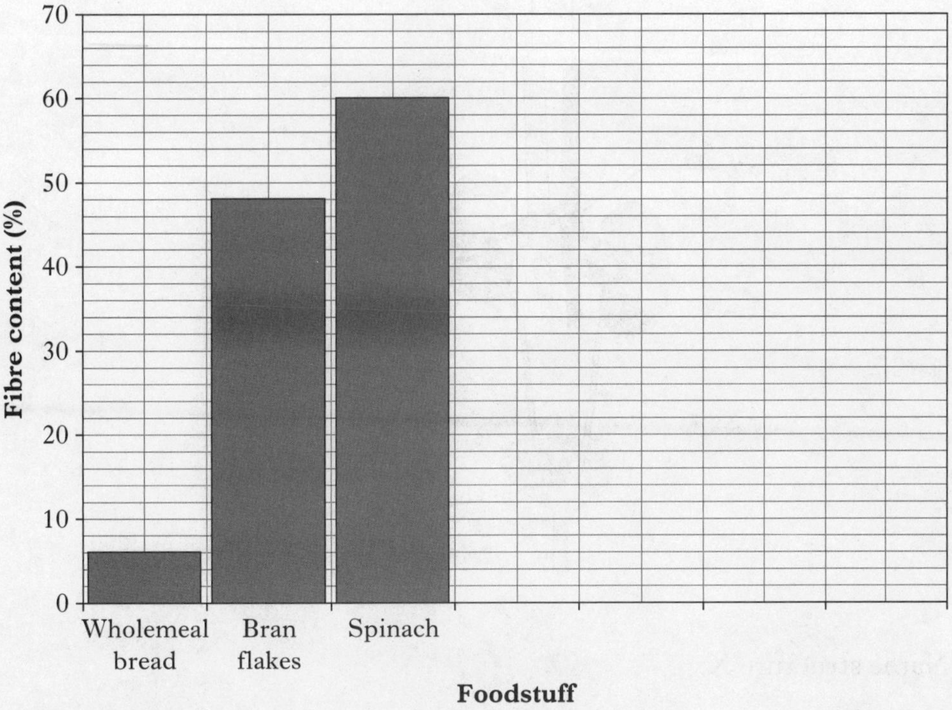

(ii) Calculate the average fibre content of the foods in the table.

_____

_____

(iii) Calculate the simple whole number ratio of fibre in bran flakes to that in wheat flakes.

Space for working

Ratio _____ : _____

Bran flakes   wheat flakes

(iv) Calculate the number of grams of fibre present in a 25g portion of spinach.

Space for working

_____ %

	KU	PS
		2
		1
		1
		1

| | KU | PS |

7. (a) The diagram below shows part of the system that regulates the body's water system.

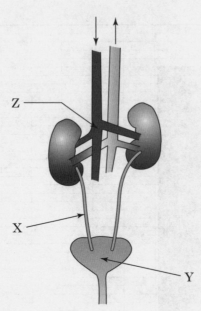

(i) Name structure X.

_____.   1

(ii) What is the function of structure Y?

_____.   1

(iii) Does blood vessel Z contain purified or unpurified blood?

_____.   1

(iv) Name the toxic waste product removed from the blood by filtration in the kidneys.

_____.   1

(b) The table below shows the daily water losses and gains of a person.

WATER LOSS (CM3)	WATER GAIN (CM3)
Sweat- 600	Food- 900
Breath-	Drink- 1100
Urine- 1200	Respiration-
Faeces- 300	TOTAL- 2400
TOTAL- 2400	

(i) Complete the table by calculating the missing totals for breath and respiration.

Space for calculation.   1

(ii) The body normally achieves water balance, how does evidence from the table support this?

(iii) Calculate the simple whole number ratio of water lost in urine to water lost from sweating.

Space for calculation.

_____

_____

URINE: SWEATING:

(iv) What percentage of the total water loss is sweat?

Space for calculation

_____%

8.

Nucleus

Cell membrane   Cytoplasm

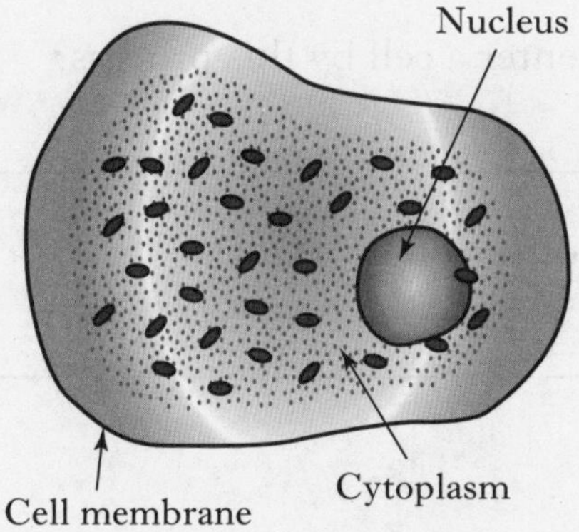

(a) The above diagram shows a picture of what kind of cell?

_____.

(b) Complete the table by ticking whether the structures are present in animal cells, plant cells or both.

STUCTURE	ANIMAL CELL	PLANT CELL	BOTH
NUCLEUS			
CYTOPLASM			
CELL WALL			
CELL MEMBRANE			
CHLOROPLAST			

9. The diagram below shows area A, magnified.

The concentration of carbon dioxide is shown by the number of dots.

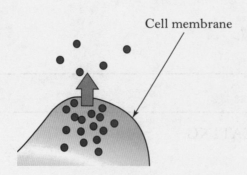

Cell membrane

(a) Name the term used to describe the type of movement shown above.

_____ 1

(b) What part of the cell controls movement of substances?

_____ 1

(c) Name two substances that enter a cell by this process?

_____ and _____ 2

(d) What name is given to the diffusion of water?

_____ 1

KU	PS

10. Egg white is a protein. Protease enzyme will digest cubes of boiled egg white.

The effect of temperature on the digestion of egg white was investigated by a student.

The results are shown in the graph below.

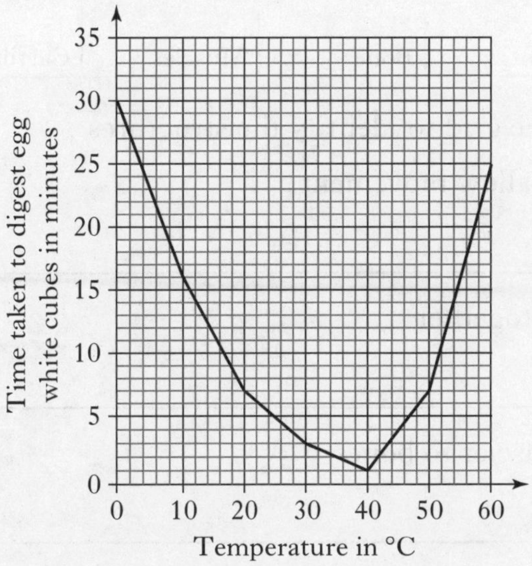

Time taken to digest egg white cubes in minutes (y-axis)

Temperature in °C (x-axis)

(a) Estimate the time taken to digest the egg white at 8°C?

_____ minutes.

(b) At what temperature did the enzyme work best?

_____°C.

(c) Why was the time taken to digest the egg white greater at 60°C than at 30°C?

_____

_____

(d) What temperature range took the greatest time to break down the egg white?

0–10°C ☐   20–30°C ☐   50–60°C ☐

KU	PS
	1
	1
1	
	1

11. (*a*) Vertebrates have a skeleton to provide support.

The grid below consists of the names of structures associated with the skeleton.

A	B	C	D	E
Tendons	Ligaments	Bones	Muscles	Cartilage

Use the letters from the grid to identify the structures

(i)   which contract to allow movement

_____   1

(ii) which hold joints together

_____   1

(iii) which attaches muscles to bones

_____   1

(*b*) The skeleton supports movement.

State one **other** function of the skeleton.

_____   1

(*c*)

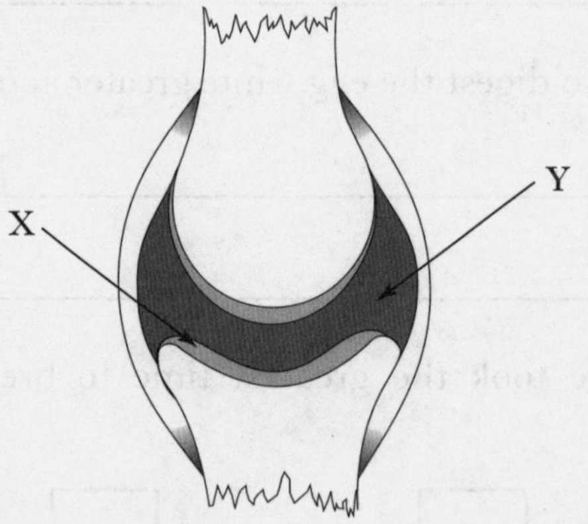

Describe the function of X?

_____

_____   1

KU | PS

12. The graph below shows the number of dead donors and the number of transplants carried out from 1996 to 2005.

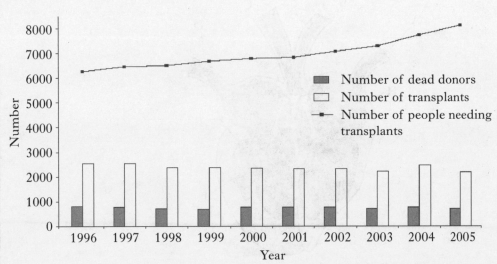

(a) The government is keen for more people to be donors.

Use information from the graph to explain why. _____

_____

_____

(b) One of the most commonly transplanted organs is the kidney.

What treatment can be used for a patient while they are waiting for a transplant if their kidneys are damaged?

_____

(c) Give one advantage of this process over having a transplant.

_____

KU	PS
	2
1	
1	

13. The diagram below shows a section of the heart of a mammal.

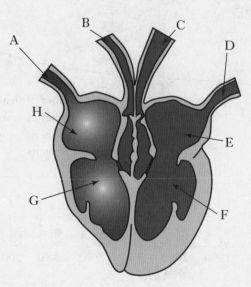

	KU	PS

Use letters from the diagram to identify the following.

(a) The vessel that takes blood from the heart to the rest of the body.

_____.   1

(b) The two ventricles.

_____ and _____.   1

(c) The vessel that takes blood from the heart to the lungs.

_____.   1

(d) What is the function of a valve?

_____   1

(e) Why is the left ventricle wall thicker than the right?

_____   1

14. The pulse rates of a trained and an untrained athlete were measured during an exercise session.

The results are shown in the table.

TIME (MINUTES)	0	1	2	3	4	5	6	7	8
PULSE RATE (BEATS PER MINUTE) UNTRAINED ATHLETE	75	90	130	170	170	150	130	110	110
PULSE RATE (BEATS PER MINUTE) TRAINED ATHLETE	65	65	85	97	115	115	105	100	80

(a) Complete the line graph of the results by

    (i) Labelling and adding an appropriate scale to the vertical axis.

    (ii) Plotting the graph of the untrained athlete's results.

The graph of the trained athlete has been plotted for you.

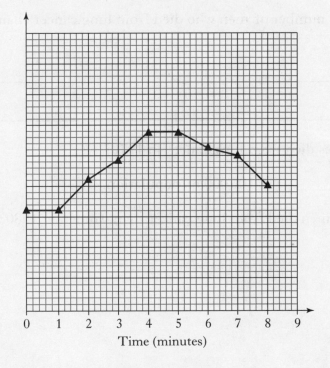

Time (minutes)

(b) What evidence shows that both athletes had not fully recovered at 8 minutes?

_____.

(c) How does the pulse rate during exercise of the trained athlete differ to that of the untrained athlete?

_____.

(d) Suggest an adaptation to the experiment that would allow the recovery time of both athletes to be measured.

_____.

KU	PS
	1
	1
	1
	1
	1

15. The graph below shows the number of people aged between 35 and 54 who died from lung cancer between 1950 and 2000.

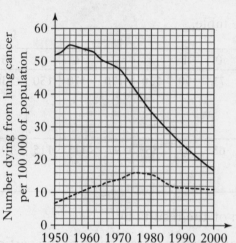

Key: —— Males aged 35–54
---- Females aged 35–54

(a) Describe how the number of men who died from lung cancer changed from 1950 to 2000.

_____

_____   **2**

(b) How many females died from lung cancer in 1970?

_____ per 100 000 of population.   **1**

(c) How many more men died from lung cancer than women in 1980?

Space for working

_____ per 100 000 of population   **1**

16. (*a*) The table below shows information about an experimental cross involving the shape of pea plant seeds.

The table shows a cross between two true-breeding parents.

GENERATION	SYMBOL	PHENOTYPES.
PARENTS	P	ROUND X WRINKLED
FIRST GENERATION		
SECOND GENERATION	F2	75%ROUND
		25%WRINKLED

(i) State the symbol for the first generation.

_____

(ii) Give the phenotype(s) of the first generation.

_____

(*b*) Decide if each of the following statements is true or false and tick the appropriate box. If the statement is false, write the correct word in the correction box to replace the word underlined in the statement.

STATEMENT	TRUE	FALSE	CORRECTION
The physical appearance of an organism is its genotype.			
Characteristics of offspring are controlled by genes.			
Sex cells are called gametes.			

(*c*) The following table contains information about the milk yield of British cows over a five year period.

YEAR	ANNUAL MILK YIELD (LITRES)
1960	3379
1961	3446
1962	3521
1963	3596
1964	3643

(i) Use the information in the table to complete the bar chart below to show the annual milk yield of cows by:

1. Labelling the vertical axis.

2. Adding a scale to the vertical axis.

3. Completing the bars.

KU: 1, 1, 3

PS: 1, 1, 1

KU | PS

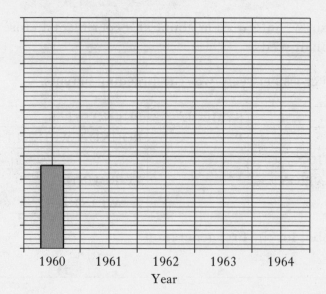

(ii) Describe the trend of milk yield between 1960 and 1965.

_____

1

(iii) Calculate the average milk yield per year from 1960 to 1965.

_____

1

17. The pictures show some products that can be made using micro-organisms.

A	B	C	D

(a) Complete the table to state what type of micro-organism is used to make the products in the pictures.

Substance	Made by
A –antibiotic	Fungus
B- wine	
C- cheese	
D- mycoprotein	fusarium

(b) The diagram shows the steps involved in inoculating an agar plate.

A. The lid of the petri dish is raised just sufficiently to allow microorganisms to be transferred from the inoculating loop to the nutrient agar

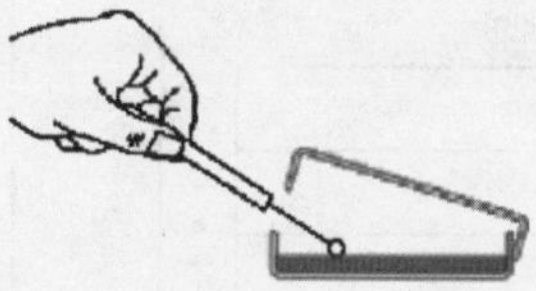

B. The petri dish with nutrient agar is sterilised by heating to 120°C

D. An inoculating loop is sterilised by heating in a Bunsen flame before and after transferring microorganisms

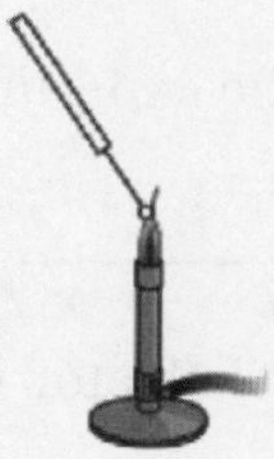

E. The cooled inoculating loop is used to collect microorganisms from a culture

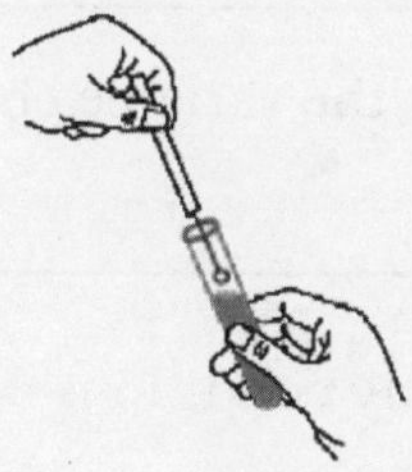

C. The lid of petri dish is sealed with adhesive tape to prevent contamination by microorganism from the air

Write the letters that label the steps in the correct order.

		E		

KU: 1

PS: 1

18.

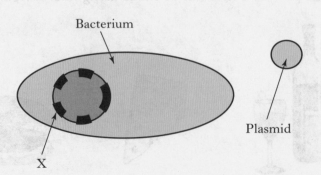

	KU	PS

(a) (i) Name structure X.   **1**

_____.

(ii) State the name of the process that allows scientists to transfer pieces of chromosome from one organism to another organism.   **1**

_____.

(b) An experiment was set up to investigate the effect of temperature on yeast activity. Glucose solution was added to live yeast cells to give each cylinder a starting volume of 50cm³. Each cylinder was kept at a different temperature. Yeast activity is shown by the height of froth of carbon dioxide bubbles made in each cylinder.

The table below shows the results after an hour.

TEMPERATURE (°C)	HEIGHT OF FOAM (CM³)
10	0
20	12
30	20
40	43
50	37
60	35

(i) What was the variable changed in the experiment?

_____.   **1**

(ii) How many conditions of the variable factor were investigated?

_____.   **1**

(iii) State one variable that would have to be kept constant.

_____.   **1**

(iv) How can the experiment be improved to increase the reliability of the results?

_____.   **1**

End of question paper.

Exam B

Biology

Standard Grade: General

Practice Papers
For SQA Exams

Exam B
General Level

Fill in these boxes:

Name of centre

Town

Forename(s)

Surname

You have 1 hour 30 minutes to complete this paper.

Try to answer all of the questions in the time allowed.

Write your answers in the spaces provided, including all of your working.

Leckie×Leckie

Scotland's leading educational publishers

		KU	PS

1. (a) The diagram below shows a food web from a marine ecosystem.

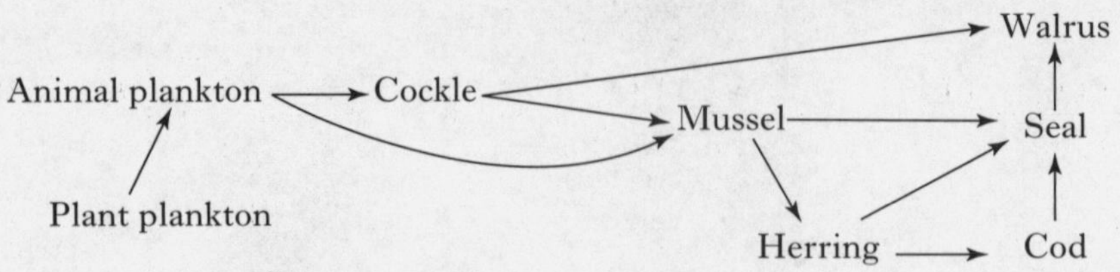

The following statements refer to the food web.

Complete the table by entering "T" when the statement is true and "F" when the statement is false.

Statement	T or F
Cockles are eaten by mussels and walruses.	
Animal plankton is not eaten by anything.	
Seals eat cod and mussels.	

PS **1**

(b) Plant plankton are producers and mussels are consumers.

Give the definition of these two terms.

PRODUCER_____ **1**

CONSUMER_____ **1**

(c) Apart from mussels, give an example of a consumer from the food web.

_____ **1**

(d) How many food chains in the food web involve cockles?

_____ **1**

(e) Give two ways in which energy can be lost from this food web.

1. _____ .

2. _____ . **2**

(f) What do the arrows in a food chain represent?

_____ **1**

2. The table below shows the results of an investigation into the distribution of dandelions in an area of grassy field and part of an oak wood.

SAMPLE SITE	1	2	3	4	5	6	7	8	9	10
ABUNDANCE OF DANDELIONS (SCORE OUT OF 25)	21	20	19	20	11	7	4	2	0	0
LIGHT INTENSITY (A=LOW H=HIGH)	H	H	H	H	G	F	E	D	C	B

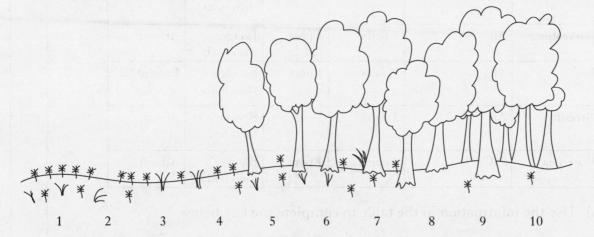

(a) Use the results to complete the bar chart of abundance of dandelions by-

(i) Adding a scale on the vertical axis.

(ii) Labelling the vertical axis.

(iii) Plotting the remaining bars.

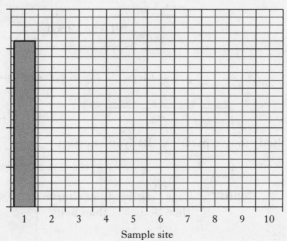

Sample site

(b) (i) Calculate the average number of dandelions present in areas 1–10.

Space for calculation

_____

(ii) Describe the relationship between the abundance of dandelions and light intensity.

_____

(iii) Light is an abiotic factor that affects distribution of organisms. Give another example of an abiotic factor affecting distribution of organisms.

_____

3. Some features of six species of soil animals are shown in the table below.

NAME	LEGS JOINED	PAIRS OF LEGS	BODY PARTS	LEG LENGTH	STRIPES
Caterpillar	Some	None	1 part	Shorter than body	Absent
Spider	All	4 pairs	2 parts	Shorter than body	Absent
Harvestman	All	4 pairs	1 part	Longer than body	Absent
Bee	All	3 pairs	1 part	Shorter than body	Present
Greenfly	All	3 pairs	1 part	Shorter than body	Absent
Waterflea	All	4 pairs	1 part	Shorter than body	Absent

(a) Use the information in the table to complete the key below.

Write the correct feature on each dotted line and the correct names in the boxes.

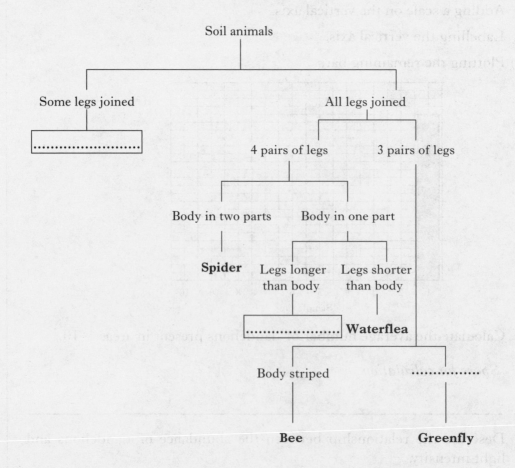

KU 1

PS 3

	KU	PS

(b) Which feature distinguishes between a bee and a greenfly?

_____ **1**

(c) What features do the waterflea and the harvestman have in common?

_____ **1**

4. (a) The diagram below shows part of a tulip flower.

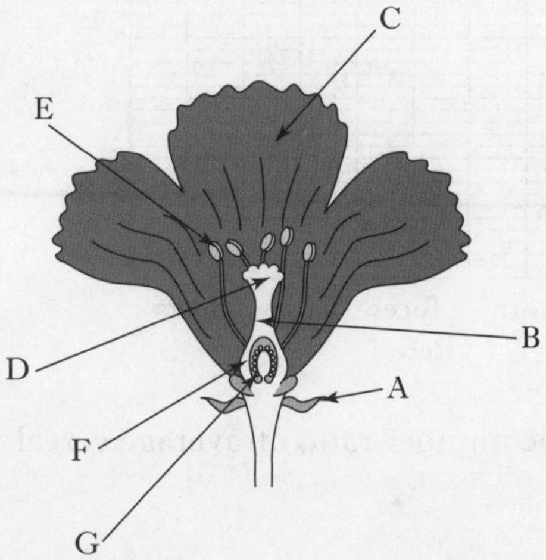

(i) Which letter indicates a structure that protects the unopened flower bud.

_____ **1**

(ii) Give the name and function of the structure labelled D.

NAME _____.

FUNCTION _____. **1**

(iii) What structure makes pollen grains?

_____ **1**

(iv) What is formed when the pollen meets the ovule?

_____ **1**

(v) What does the ovary become after fertilisation?

_____ **1**

(b) The table below shows the average mass of cereal produced by different types of plant.

CEREAL	MILLIONS OF METRIC TONNES
Wheat	300
Corn	227
Rice	160
Barley	100
Oats	46

Complete the bars on the graph by-

(i) Adding a suitable scale to the Y-axis.

(ii) Adding a label to the Y-axis.

(iii) Completing the bars on the graph.

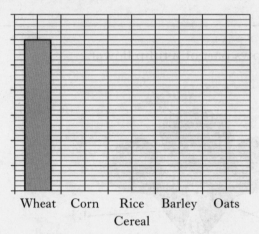

Wheat   Corn   Rice   Barley   Oats

Cereal

(iv) Calculate the simple whole number ratio of average cereal yield in wheat to that of barley.

Space for calculation

Wheat : Barley

(v) How many times greater than barley is the weight of wheat?

KU	PS
	1
	1
	1
	1
	1

	KU	PS

5. (*a*) The diagram below shows the skulls of two mammals.

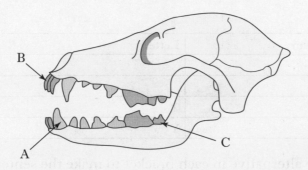

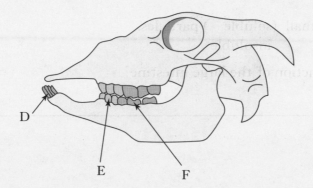

Use letters from the diagram to identify the following teeth.

(i) Canines _____ **1**

(ii) A tooth used to grind and crush plant material _____ **1**

(iii) A tooth used for piercing and holding prey _____ **1**

(*b*) The diagram below shows the human digestive system.

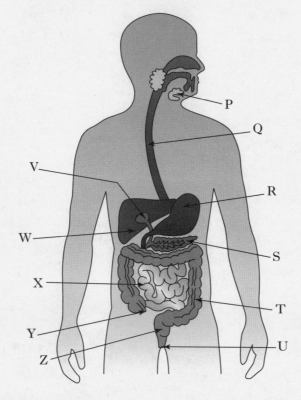

	KU	PS

(i) Complete the table below to identify the following parts of the digestive system.

Part of digestive system	Letter
Salivary glands	
Stomach	
	T
	Y

KU **2**

(ii) <u>Underline</u> one alternative in each bracket to make the sentence correct.

Digestion is the process by which mammals break down large (soluble)

insoluble

particles into small (soluble) particles.

insoluble

KU **2**

(iii) What is the function of the large intestine?

_____

KU **1**

KU | PS

6. The following experiment shows the effect of an unknown enzyme on the breakdown of starch.

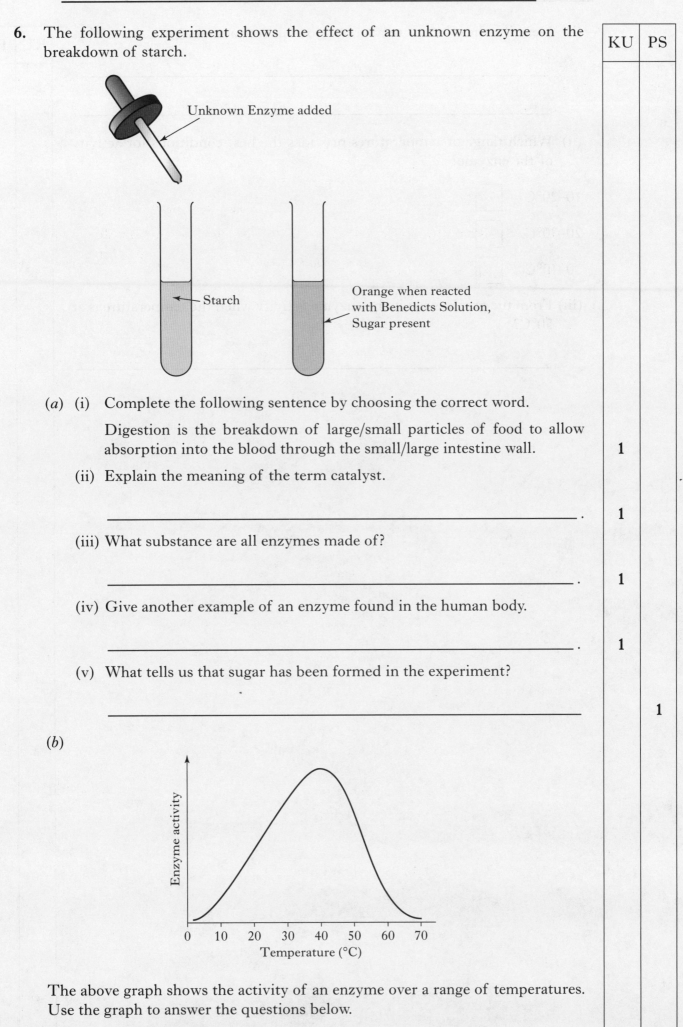

Unknown Enzyme added

Starch

Orange when reacted with Benedicts Solution, Sugar present

(a) (i) Complete the following sentence by choosing the correct word.

Digestion is the breakdown of large/small particles of food to allow absorption into the blood through the small/large intestine wall.

1

(ii) Explain the meaning of the term catalyst.

_____.

1

(iii) What substance are all enzymes made of?

_____.

1

(iv) Give another example of an enzyme found in the human body.

_____.

1

(v) What tells us that sugar has been formed in the experiment?

_____

1

(b)

Enzyme activity

0 10 20 30 40 50 60 70
Temperature (°C)

The above graph shows the activity of an enzyme over a range of temperatures. Use the graph to answer the questions below.

	KU	PS

(i) Describe the effect of temperature on enzyme activity.

_____

_____

2

(ii) Which range of temperatures provides the best conditions for activity of the enzyme?

10–20°C ☐

20–30°C ☐

30–40°C ☐

1

(iii) From the graph, what is the enzyme activity when the temperature is at 50°C?

_____

1

7. Maggots move away from light. Increasing light intensity can affect their rate of movement. This was investigated using the apparatus shown below.

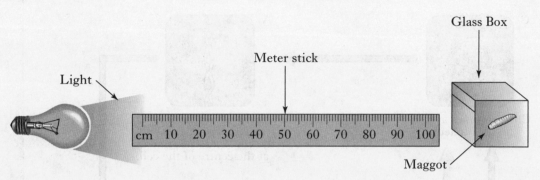

The time taken for the maggot to move 15cm was recorded when the lamp was moved along the metre stick. This was repeated with another two maggots. The results are shown in the table.

Distance between lamp and box(cm)	Time taken to move 15cm			
	Maggot 1	Maggot 2	Maggot 3	Average
100	35	33	31	33
50	24	30	30	
25	21	18	15	18

(a) Complete the table with the average time for the maggots to move 15cm when the lamp is 50cm away.

Space for calculation

(b) Underline one option to complete the following sentences:
As the distance between the lamp and the glass box decreases, the light intensity
Increases
Stays the same
Decreases

As the light intensity decreases, the time taken for the maggot to move 15 cm
Increases
Stays the same
Decreases

(c) What would happen to the rate of movement if the maggots were placed in the dark?

_____

(d) How could the reliability of the results be improved?

_____

KU | PS

1

1

1

1

1

8. (a) The following diagrams show four stages of mitosis but not in the correct order.

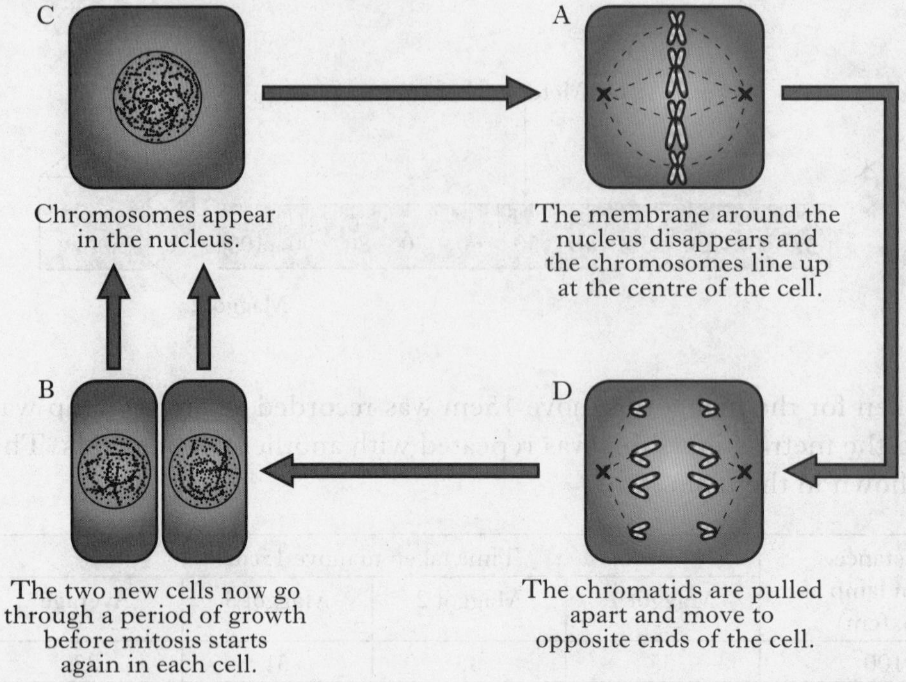

C

Chromosomes appear
in the nucleus.

A

The membrane around the
nucleus disappears and
the chromosomes line up
at the centre of the cell.

B

The two new cells now go
through a period of growth
before mitosis starts
again in each cell.

D

The chromatids are pulled
apart and move to
opposite ends of the cell.

(i) Arrange the letters from the diagrams to put the stages into the correct order.

1ST STAGE _____.

2ND STAGE _____.

3RD STAGE _____.

4TH STAGE _____.

(ii) Complete the following sentence by underlining the correct option in each group.

Mitosis occurs in the Cytoplasm   of a cell and it increases
                       Nucleus   decreases
                       Cell membrane   does not change

the number of cells present in an organism.

KU	PS

2

2

(b) An investigation was carried out looking at the growth of bacteria. The results are shown below.

TIME (SECONDS)	0	30	60	90	120	150
NUMBER OF BACTERIA (Thousands per mm³)	3	12	20	24	48	94

(i) Complete the line graph below by-

1. Adding a suitable scale to the y-axis.

2. Adding a label to the y-axis.

3. Plotting the graph.

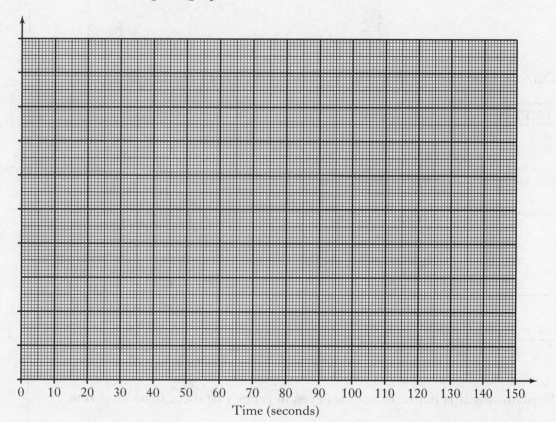

Time (seconds)

(ii) Give the ratio of bacteria present at 30 seconds to that of bacteria present at 90 seconds.

Space for calculation.

30 seconds: 90 seconds:

(iii) From the graph give the number of bacteria when the time is 100 seconds?

_____.

KU | PS

3

1

1

9. The diagram below shows part of the human eye.

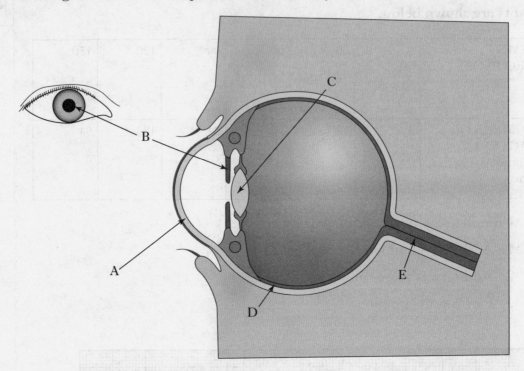

(a) Complete the table to show the name and function of the labelled parts.

LETTER	NAME	FUNCTION
A	Cornea	
	Optic nerve	
D		Light sensitive layer.
C		Flexible structure, focussing light onto retina.
	Iris	Coloured part of the eye.

3

(b) Explain why it is better to use two eyes to judge distances instead of one.

_____.

1

10. The diagram below shows part of the human lungs.

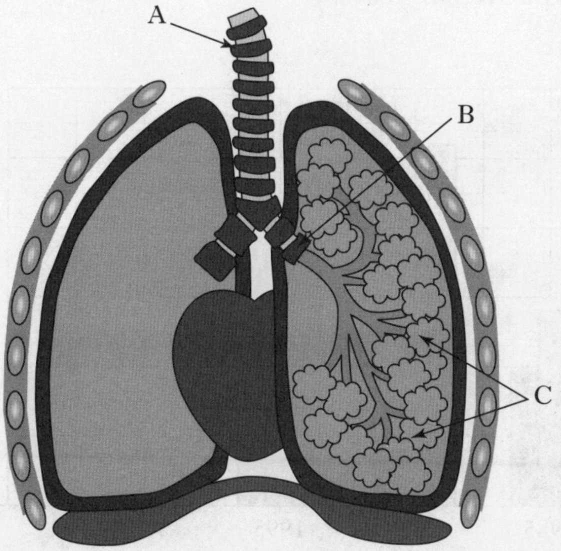

(a) Complete the table below to show the name and functions of the labelled parts.

LETTER	NAME	FUNCTION
C		Air sacs allowing oxygen to pass to the lungs.
	Bronchi	Two divisions of windpipe.
A		Tube from mouth to bronchi

(b) Use lines to link the parts of the skeleton to each of the organs they protect.

PART OF SKELETON   ORGANS.

Rib cage   Brain

Vertebrae   Heart

Skull   Spinal cord

11. The bar chart below shows how much phosphate entered a lake each day from different sources in three particular years.

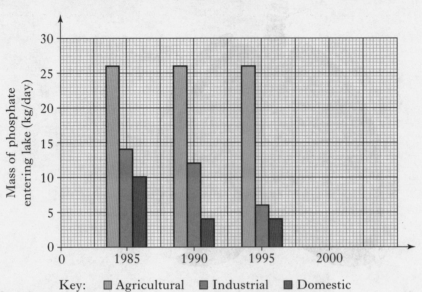

Key: ■ Agricultural ■ Industrial ■ Domestic

(a) Which source of phosphate remained the same?

_____

(b) What mass of phosphate entered the lake each day in discharges from Industry in 1990?

_____

(c) What was the difference between the domestic discharge in 1985 and 1995?

_____

(d) The figures for the year 2000 are as follows

AGRICULTURAL- 26

INDUSTRIAL- 4

DOMESTIC- 2

Complete the bar graph to show this information.

KU	PS
	1
	1
	1
	2

	KU	PS

12. (*a*) (i) Complete the diagram to show the sex chromosomes present in each of the cells.

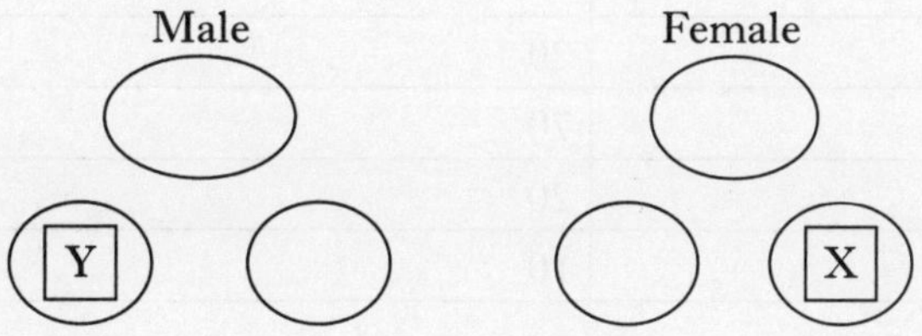

Male   Female

(ii) How many sets of chromosomes are present in an egg cell?

_____.

(iii) What is the name given to sex cells?

_____.

(iv) What name is given to the process when the male and female sex cells join together?

_____.

(v) What term is used to describe the physical appearance of an individual?

_____

(*b*) Complete the following sentences by underlining the correct word in each bracket.

A $\begin{pmatrix} \text{fertile} \\ \text{Species} \\ \text{Clone} \end{pmatrix}$ is a group of living things that can interbreed and

produce $\begin{pmatrix} \text{fertile} \\ \text{sterile} \\ \text{different} \end{pmatrix}$ offspring.

KU marks: 2, 1, 1, 1, 1, 2

13. The shoe sizes of 120 school pupils are shown in the table below.

SHOE SIZE	NUMBER OF SCHOOL PUPILS
4	20
5	70
6	20
7	10

(a) Complete the pie chart below to show this information.

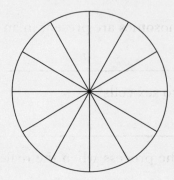

(b) What type of variation is shown by shoe size?

_____.

(c) What percentage of pupils have size 5 feet?

Space for calculation

_____.

KU | PS

2

1

1

14.

	KU	PS

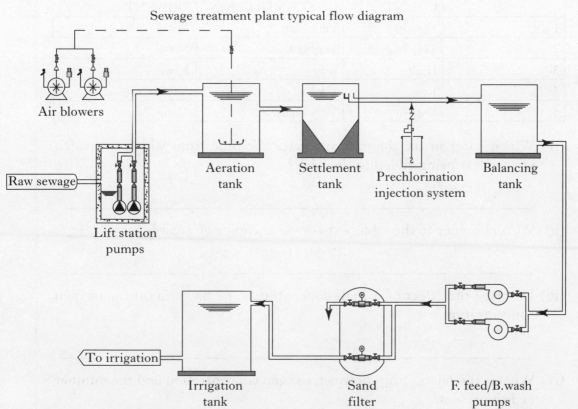

Sewage treatment plant typical flow diagram

Air blowers

Raw sewage

Lift station pumps

Aeration tank

Settlement tank

Prechlorination injection system

Balancing tank

To irrigation

Irrigation tank

Sand filter

F. feed/B.wash pumps

(*a*) (i) Name the gas produced in the sludge tank which can be used as a fuel.

_____ .

1

(ii) State one advantage of using fuels obtained by fermentation rather than fossil fuels.

_____

1

(*b*) The following diagram shows the results of a water sample experiment looking at the effect of sewage on a river.

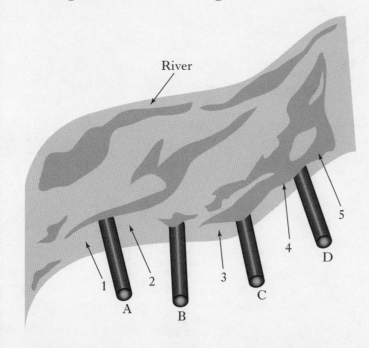

River

1

2

3

4

5

A

B

C

D

SAMPLE POINT	BACTERIAL COUNT	OXYGEN CONCENTRATION	NUMBER OF FISH PRESENT
1	Very low	Very high	Many
2	Very high	Very low	None
3	High	Low	None
4	Low	High	A few
5	Very low	Very high	Many

(i) Which letter in the diagram indicates the pipe from which untreated sewage was being added to the river?

_____.

(ii) With reference to the table, explain your choice of answer.

_____.

(iii) Describe the effect of a high concentration of bacteria on the oxygen concentration.

_____.

(iv) What is the relationship between oxygen concentration and the number of fish present?

_____

	KU	PS

15. The following picture shows the principal precautions and sterile techniques used while carrying out laboratory work.

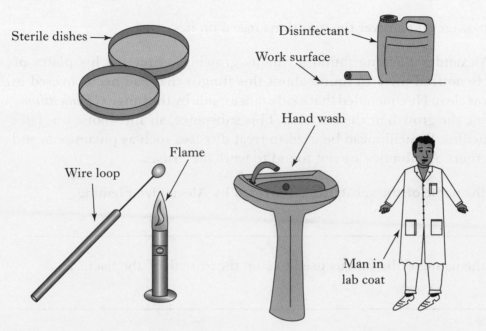

Fill in the following table to explain the reason for each method being used.

STERILE TECHNIQUE	REASON
Sterile Petri dishes kept closed.	
Wire loop flamed before use.	
Work surface cleaned with disinfectant.	
Hands washed.	
Lab coat worn.	

5

16. The following passage contains information about Alexander Fleming and his scientific discovery.

Read the passage and answer the questions based on it.

In 1928 Alexander Fleming found a fungus growing on one of his plates of bacteria. He noticed the area round about this fungus that had been covered in bacteria, was clear. He concluded that a substance made by the fungus (*Penicillium*) was stopping the growth of the bacteria. This substance, an anti-biotic was later called Penicillin. Penicillin can be used to treat diseases such as pneumonia and chest infections. Antibiotics are not found to work on viruses.

(*a*) Give the name of the substance discovered by Alexander Fleming.

_____.

(*b*) Give the name of the fungus used to stop the growth of the bacteria.

_____.

(*c*) What is the name of the first antibiotic?

_____.

(*d*) Give an example of a disease that the product discovered can be used to treat.

_____.

(*e*) What do antibiotics not work on?

_____.

KU	PS
	1
	1
	1
	1
	1

[End of question paper]

Exam C

Biology

Standard Grade: General

Practice Papers
For SQA Exams

Exam C
General Level

Fill in these boxes:

Name of centre

Town

Forename(s)

Surname

You have 1 hour 30 minutes to complete this paper.

Try to answer all of the questions in the time allowed.

Write your answers in the spaces provided, including all of your working.

Leckie×Leckie

Scotland's leading educational publishers

1. The following grid contains some terms relating to the biosphere topic.

A Habitat	B Producer	C Consumer	D Community
E Population	F Decomposer	G Ecosystem	H Light intensity

Use **letters** from the grid to complete the following statements.

(a) The place where an organism lives is its _____.

(b) An organism which makes its own food is _____.

(c) An ecosystem is made up of _____ and _____.

(d) An example of an abiotic factor is _____.

2. The diagram shows part of a food web for a small pond.

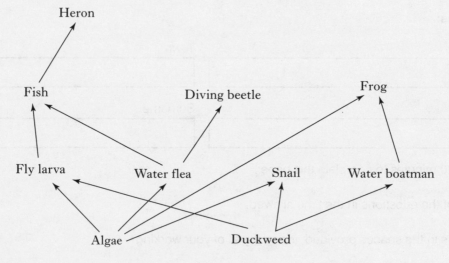

From the web, choose a predator and name its prey.

predator : _____

prey : _____

(b) Which organism feeds on both animals and plants?

_____

(c) Name a producer.

_____

(d) How many food chains involve fly larva?

_____

KU: 1, 1, 1, 1, 1, 1

PS: 1, 1

3. The lynx is a wild cat that lives in Canada. The table shows the number of lynx trapped in a part of Canada in certain years.

Year	Number of lynx in thousands
1918	45
1920	25
1922	10
1924	20
1926	40
1928	50

The snowshoe hare is another wild animal found in Canada. The graph shows the number of snowshoe hares trapped in the same years. The lynx eats the snowshoe hare.

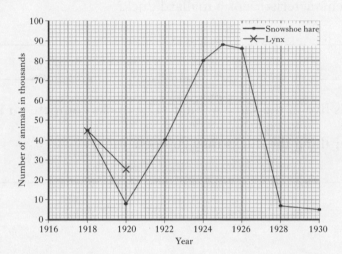

(a) Draw a graph of the data in the table. The first two points have been plotted.

2

(b) From the graph, predict how many snowshoe hare would have been trapped in 1921?

_____thousand

1

(c) From your graph, predict how many lynx were trapped in 1925.

_____ thousand

1

(d) Use the information to answer the following.

   (i) What would you expect to happen to the number of lynx trapped in 1930?

   Underline your answer.

   Rise   **Fall**   **Stay the same**

1

  (ii) Give a reason for your answer to part (i).

_____

_____

1

KU   PS

4. The diagram shows a key used to name four different ducks.

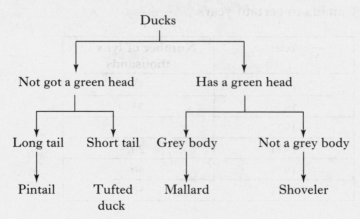

(a) Write down two characteristics of a mallard duck?

1. _____

2. _____

(b) All of these ducks live in the same habitat. The ducks compete for space.

Suggest one other thing the ducks may compete for.

_____

5. The table below shows the mass of some pollutants entering the air in Scotland.

Pollutant	% of total pollution
Carbon monoxide	50
Sulphur dioxide	20
Hydrocarbons	10
Dust	15
Nitrogen oxides	5
Total	100

(a) Use the information from the table to complete the pie chart below.

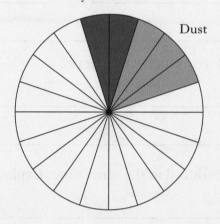

Hydrocarbons

Dust

(b) If the total mass of pollution was 300 units, how many units were caused by dust?

Space for calculation.

Answer _____ units

(c) In 1980, the concentration of smoke in the air was 45 units/m^2

In 2000, it was 15 units/m^2.

How many times greater were smoke concentrations in 1980 than in 2000?

Space for calculation

Answer _____times

KU | PS

3

1

1

6. The diagram shows a sperm cell and an egg cell.

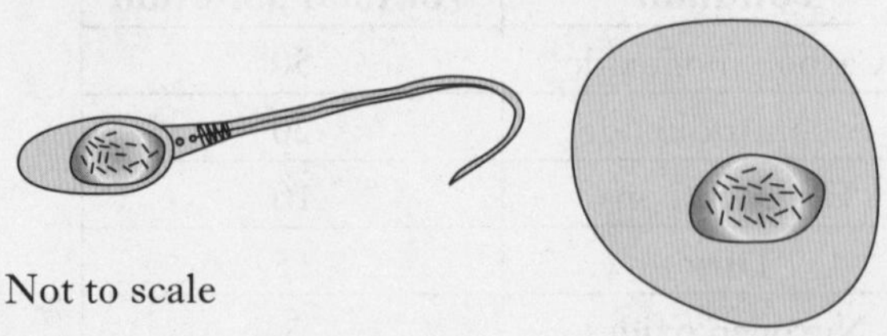

KU	PS

Sperm   **Egg**

Not to scale

(a) (i) From the diagrams alone, describe one way in which the sperm cell is different from the egg cell.

_____

_____   **1**

(ii) Explain how this difference helps the sperm.

_____

_____   **1**

(b) Lots of sperm cells are produced at the same time. Explain why.

_____

_____   **1**

(c) Kangaroos have 12 chromosomes in each skin cell.

How many chromosomes would be in a sperm cell from a kangaroo?

_____   **1**

(d) Underline one word in the brackets to complete the following sentences correctly.

The female sex cells are made in the $\begin{pmatrix} \text{OVARY} \\ \text{OVULES} \\ \text{TESTES} \end{pmatrix}$ and are called $\begin{pmatrix} \text{SPERM} \\ \text{OVULES} \\ \text{OVIDUCT} \end{pmatrix}$ **1**

The male sex cells are made in the $\begin{pmatrix} \text{TESTES} \\ \text{OVARY} \\ \text{SPERM} \end{pmatrix}$ and are called $\begin{pmatrix} \text{SPERM} \\ \text{OVULES} \\ \text{OVIDUCT} \end{pmatrix}$ **1** (*e*) The diagrams below show the human male and female reproductive systems.

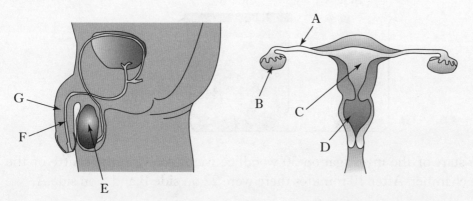

Complete the table below by adding the correct letter, name and function of the parts.

LETTER	NAME	FUNCTION
B		Where eggs are made
	Uterus	Where the foetus would develop.
E		Where sperm are made
G	Penis	

KU | **PS**

4

	KU	PS

7. The following experiment was set up to investigate the behaviour of woodlice.

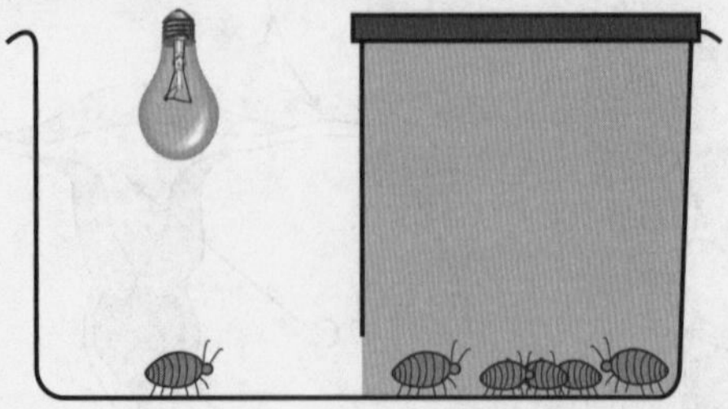

Side A   Side B

At the start of the investigation 30 woodlice were placed in the centre of the choice chamber. After 10 minutes there were 22 on side B and 8 on side A.

(a) What environmental factor is being investigated?

_____.   1

(b) Describe the response of the woodlice in the investigation.

_____.

_____   1

(c) Why were 30 woodlice used instead of one?

_____.   1

(d) Why were the woodlice left for 10 minutes before the results were taken?

_____.

_____   1

(e) Give another example of a situation where an environmental factor has an effect on a named animal's behaviour.

_____.

_____   1

8. Read the passage below and answer the questions based on it.

KU	PS

MIGRATING BIRDS

The fantastic annual migrations that birds make between their breeding and wintering grounds is one of the wonders of our natural world. Most of the world's 29 or so species of geese are no strangers to migration, and some routinely accomplish amazing feats. In Asia, Bar-headed Geese (*Anser indicus*) regularly migrate over the Himalayan Mountains, even over Mount Everest at an altitude of 30,750 feet (9375 m) where the air is thin and the temperatures drop to minus 60 degrees F.

Migrating birds, especially waterfowl, follow broad but well defined migration routes called flyways or migration corridors. There are four primary corridors in North America. From east to west, they are the Atlantic, Mississippi, Central and Pacific flyways.

Many species of geese and other waterfowl breed in the far northern reaches of North America, and begin their journey south following well defined geographical features like coastlines, rivers and mountain ranges.

Snow Geese breed in the Arctic Tundra and winter in farmlands, lakes and coastal areas in the American south, southwest and east coast. These attractive geese occur only in North America, and make an annual round trip journey of more than 5,000 miles at speeds of 50 mph or more. Seen in flight, adults are white with jet black wing tips.

(*a*) Where are the Bar-headed geese found to migrate?

_____   **1**

(*b*) Give two of the four primary corridors that birds follow on migrations.

_____   **1**

(*c*) Give two species that breed in the northern parts of North America.

_____   **1**

(*d*) Where do snow geese breed?

_____   **1**

(*e*) What is the distance of the snow geese migration each year?

_____   **1**

(*f*) Describe the appearance of the adult snow geese in flight.

_____   **1**

		KU	PS

9. Before 1977, otters were in danger of becoming extinct.

The table below shows numbers of otters from 1978 to 1993.

Country	Number of places surveyed	Number of places where otters were found		
		1978	1985	1993
England	2940	170	284	687
Wales	1008	207	393	529
Scotland	2650	1511	1717	2211
Great Britain (Total)	6598	1888	2394	3427

(a) (i) How has the number of places where otters were found changed between 1978 and 1993?

Underline the correct answer.

decreased   **stayed the same**   **increased**

                                                                 1

(ii) In 1985, which country had the fewest places where otters were found?

Underline the correct answer.

England   **Scotland**   **Wales**

    1

(b) What percentage of the total number of places that otters were found in 1978 were found in England?

Space for calculation

Answer _____ %

    1

(c) What was the percentage increase in the number of otters found in Scotland between 1985 and 1993?

Space for calculation

Answer _____ %

    1

(d) How many more otters were found in Wales in 1993 than 1978?

Space for calculation

Answer _____

    1

(e) Predict what will happen to the number of otters in Scotland after 1993?

Underline the correct answer.

INCREASE   DECREASE   STAY THE SAME

    1

10. The table below shows where different substances enter and leave the blood.

Substance	Where substance enters the blood	Where substance leaves the blood
Oxygen		Body tissues
	Body tissues	Lungs
Digested food		Body tissues

(a) Complete the table by filling in the missing words.

Substances such as oxygen can enter and leave cells by diffusion.

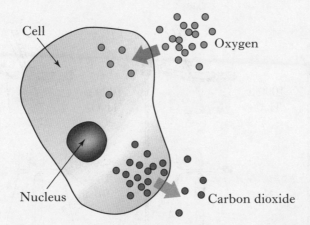

5 cells are shown below and the dots show the concentration of oxygen.

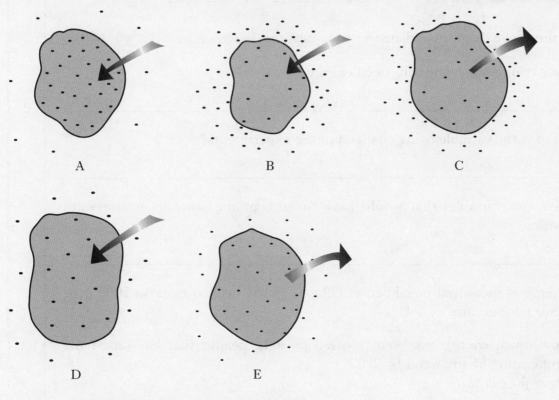

(b) Which diagram shows the diffusion of oxygen?

_____

(c) Name another substance that enters a cell by diffusion.

_____

KU | PS

2

1

1

	KU	PS

(d) What part of the cell controls what enters and leaves the cell?

_____   1

(e) What is osmosis?

_____

_____   1

11. An experiment was set up looking at energy release in cells.

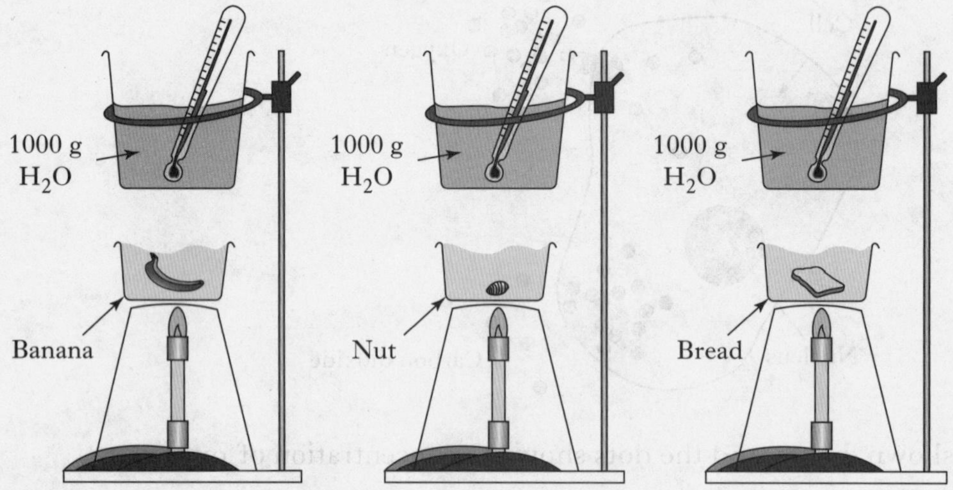

Study the above diagram and answer the questions based on it.

(a) Give two reasons why cells need energy.

_____ AND_____.   2

(b) What is the variable being changed in the experiments?

_____.   1

(c) Give two variables that would have to be kept the same to ensure valid results.

_____

_____   1

(d) Energy is measured in kilojoules (kJ), 4.2 kJ is needed to raise 1000 g of water temperature by 1°C.

How much energy has been released by a 1g peanut that has raised the temperature of the water by 2°C?
Space for calculation.

Answer _____ kJ   1

(e) What do cells need to release energy from food during aerobic respiration?

_____   1

	KU	PS

12. The diagram below shows a section through a seed.

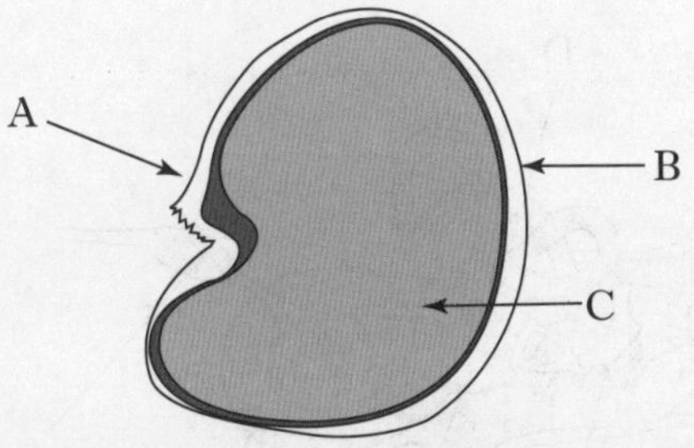

(a) Name the parts labelled A and B.

A- _____.

B- _____. **2**

(b) What part of the seed provides the growing embryo with food?

_____. **1**

(c) What is the function of part B?

_____. **1**

13. Below is a flower that is wind pollinated.

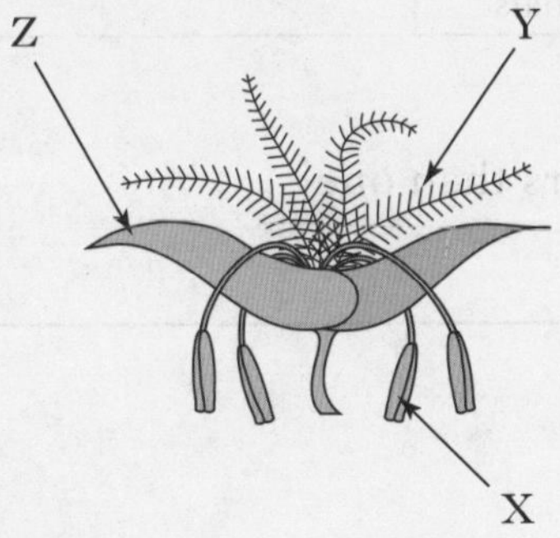

(a) Name structures X and Y.

X - _____

Y - _____ **2**

(b) Give the function of structure Y

_____

_____

_____ **1**

(c) Give the name of structure Z

_____

_____ **1**

14. Below is a diagram of the human ear.

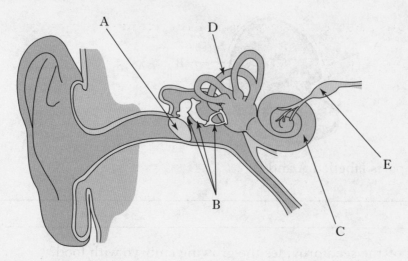

(a) Complete the table below.

Letter	Name of part	Function
A	Ear drum	Passes vibrations to middle ear
	Middle ear bones	Amplify and transmit vibrations
C		Convert vibrations into nerve impulses
D		
E	Semi circular canals	

(b) Why is it better to use two ears than one?

_____

KU	PS

3

1

15. Read the following passage and answer the questions that follow.

KU | PS

Scientists find new species of 'Hobbit' Human

The remains of a human have been discovered on an Indonesian island. The skeleton was only 1 metre tall and is 18000 years old.

The pelvis showed that the skeleton was a woman. Her teeth were worn and her skull bones fused together suggesting an adult of around 30 years old.

Nearby were found remains of stone tools, charred wood and roasted animals. These suggest that the woman was intelligent, cooked food and might even have built rafts and used language.

The 'little' humans may be descendants of a Homo erectus population that became isolated.

'Various factors such as isolation, poor resources and few predators have led to a small body size by natural selection', said one leading scientist.

(a) What evidence suggests that the woman was around 30 years old?

_____

_____   1

(b) What did the charred wood that was found nearby suggest?

_____   1

(c) What factors have led to a small body size?

_____

_____   1

(d) What were the 'little' humans descendants of?

_____   1

(e) How old was the skeleton found?

_____   1

16. (*a*) Using lines, match the components of blood to their description

COMPONENT   DESCRIPTION

Red blood cell   Watery, yellow liquid that
carries dissolved substances.

Plasma   Carries oxygen.

(*b*) The table below shows the lactic acid concentration of the blood of an athlete measured over time.

TIME (MINUTES)	LACTIC ACID CONCENTRATION OF BLOOD (mg/100 cm³)
0	10
2	10
4	22
6	38
8	41
10	43
12	46

Use the information in the table to complete the line graph to show the lactic acid concentration of blood over time by:

(i) Labelling the vertical axis

(ii) Adding a scale to the vertical axis

(iii) Completing the graph.

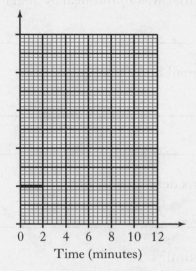

Time (minutes)

(iv) Describe the effect of increasing time on the lactic acid concentration of the athlete.

_____

(v) Calculate the average lactic acid concentration of the athlete over the 12 minute period.

_____

KU   PS

2

1

1

1

2

1

	KU	PS

17. The statements below refer to statements about inheritance.

(a) Use lines to match up each term to the right description.

TERM   DEFINITION

Phenotype   Always expressed if present

Genotype   Physical appearance

Dominant   Genetic makeup of an organism

KU: 2

(b) The bar chart below gives information about the blood groups of members of a town.

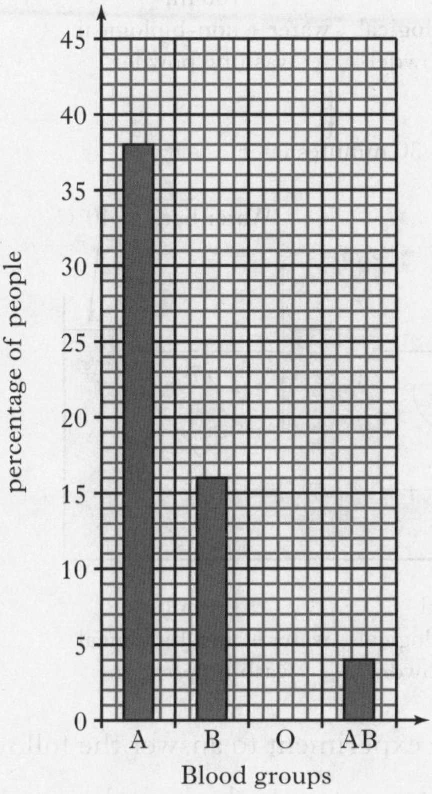

(i) What percentage of people have blood group B?

_____

PS: 1

(ii) Give the whole number ratio of people with blood group B to that of people with blood group AB.

Space for calculation.

B : AB _____ : _____

PS: 1

(iii) 42% of people have blood group O- plot this bar on the graph.

PS: 1

18. The diagram below shows an investigation onto the effect of different washing powders on stains.

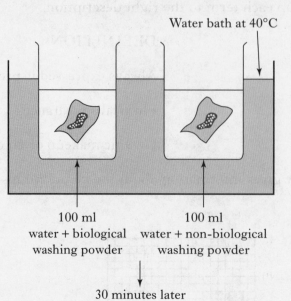

Water bath at 40°C

100 ml
water + biological
washing powder

100 ml
water + non-biological
washing powder

30 minutes later

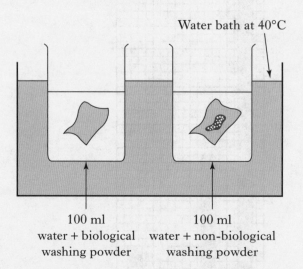

Water bath at 40°C

100 ml
water + biological
washing powder

100 ml
water + non-biological
washing powder

Use the information about the experiment to answer the following questions.

(*a*) Which of the two detergents, non-biological or biological has proved best at removing the stain?

_____

(*b*) What variable has been investigated in the two experiments?

_____

(*c*) What two variables must be kept the same to ensure reliable results?

_____

(*d*) What is the difference between biological and non-biological washing powders?

_____

KU	PS
	1
	1
	1
1	

19. The following diagram describes some of the stages involved in transferring a gene from a human chromosome into a bacterial cell.

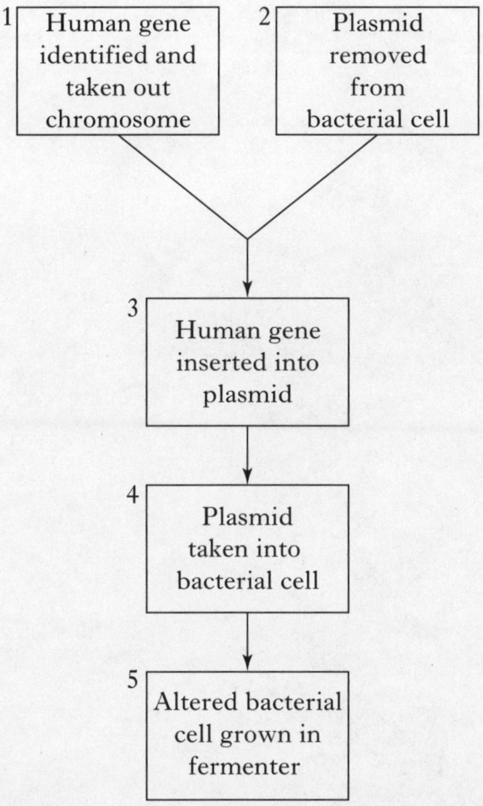

```
1 ┌──────────────┐      2 ┌──────────────┐
  │ Human gene   │        │   Plasmid    │
  │ identified and│        │  removed     │
  │ taken out    │        │   from       │
  │ chromosome   │        │ bacterial cell│
  └──────┬───────┘        └──────┬───────┘
         └───────────┬───────────┘
                     ▼
          3 ┌──────────────┐
            │  Human gene  │
            │ inserted into│
            │   plasmid    │
            └──────┬───────┘
                   ▼
          4 ┌──────────────┐
            │   Plasmid    │
            │  taken into  │
            │ bacterial cell│
            └──────┬───────┘
                   ▼
          5 ┌──────────────┐
            │Altered bacterial│
            │ cell grown in│
            │  fermenter   │
            └──────────────┘
```

(a) What name is given to this process?

_____ | 1

(b) Give an example of a product that can be made by bacteria as a result of this procedure.

PRODUCT-_____. | 1

USE-_____. | 1

(c) What type of reproduction is involved during the growth of the bacteria?

_____ | 1

(d) State one product made using bacteria.

_____ | 1

[End of question paper]

19. The following diagram describes some of the stages involved to transfer a gene from a human chromosome into a bacterial cell.

(a) What name is given to this process?

_____

(b) Give an example of a product that can be made by bacteria treated with this procedure.

PRODUCT _____

USE: _____

(c) What type of reproduction is involved during the growth of the bacteria?

_____

(d) State one product made using bacteria.

_____

[End of question paper]

Exam D

Biology

Standard Grade: General

Practice Papers
For SQA Exams

Exam D
General Level

Fill in these boxes:

Name of centre

Town

Forename(s)

Surname

You have 1 hour 30 minutes to complete this paper.

Try to answer all of the questions in the time allowed.

Write your answers in the spaces provided, including all of your working.

1. Two students estimate how many bluebell plants are in a wood. They use quadrats to collect the data.

Each quadrat covers 1 m².

	Mike's results	Linda's results
total area of wood	5000 m²	5000 m²
number of quadrats	20	10
average number of bluebells in each 1 m² quadrat	9	20
estimated number of bluebells in the wood	45 000	

(a) Complete the table to calculate the estimated number of bluebells using Linda's results.

Space for calculation

1

(b) Which student's results were more reliable?
Give a reason for your answer.

Student _____

Reason _____

_____

1

(c) What piece of equipment could possibly be used to estimate all insect types?

_____

1

(d) Light is an example of an abiotic factor, name another.

_____

1

KU PS

	KU	PS

2. Below is a food web from the African grasslands.

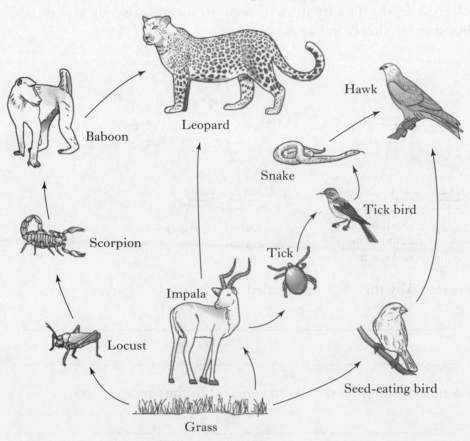

(a) Name an organism from the food web that is a :-

(i) consumer _____   1

(ii) producer _____   1

(b) Only using information from the food web, complete the food chain.

……….. ⟶ ……….. ⟶ ……….. ⟶ ……….. ⟶ leopard   1

(c) What do the arrows in a food web represent?

_____

_____   1

3. Two bags A and B were used to investigate the effect of sulphur dioxide on the germination of cress seeds. The mixture of sodium metabisulphide and water released sulphur dioxide slowly in bag A.

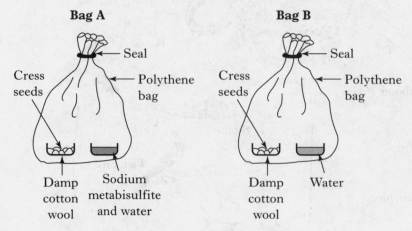

(a) Give one reason why the bags were sealed.

_____

_____ 1

(b) Give two conditions that must be kept constant in the investigation.

1 _____

2 _____ 2

(c) The sulphur dioxide spread through the bag to the cress seeds.

Which word best describes what happened to it?

Tick one box only.

Diffusion ⬚

Osmosis ⬚

Evaporation ⬚

Denaturing ⬚ 1

(d) The table shows the result of the investigation.

	Bag A	Bag B
Number of seeds	20	20
Number of seeds germinated	8	15

(i) What percentage of seeds germinated in bag B?

Space for working

_____ %

(ii) What effect did sulphur dioxide have on the germination of cress seeds?

_____

_____

(e) For a seed to germinate, oxygen is required.

Name two other factors necessary for germination to take place.

1. _____

2. _____

KU | PS

1

1

2

	KU	PS

4. The graph shows how the concentration of sugar in the leaves of a plant varies during a summer's day.

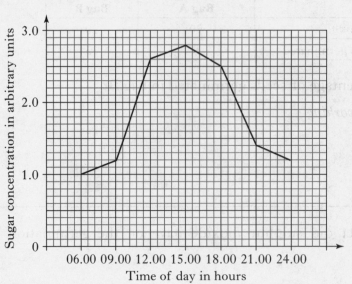

(a) Give the time when the sugar concentration is at its highest.

_____

(b) Some of this sugar will be used for respiration. How will the rest of the sugar be stored?

_____

_____

(c) In which 3 hour period did the sugar concentration increase the most?

_____

(d) Through which structures in a plant does food travel?

_____

KU / PS markings in margin: (a) PS 1, (b) KU 1, (c) PS 1, (d) KU 1

5. Four leaves were removed from the same plant. Petroleum jelly (a waterproofing agent) was spread onto some of the leaves, as follows:

Leaf A: on both surfaces

Leaf B: on the lower surface only

Leaf C: on the upper surface only

Leaf D: none applied

Each leaf was then placed in a separate beaker, as shown in the diagram.

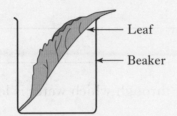

Each beaker was weighed at intervals. The results are shown in the graph below.

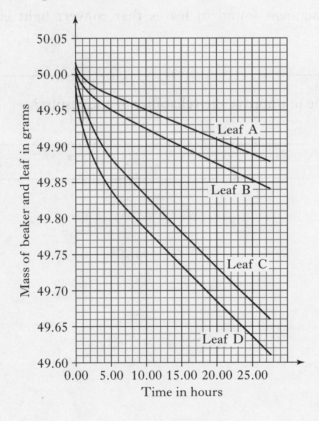

	KU	PS

(a) Using information from the graph, answer the following questions.

(i) Which surface (upper or lower) loses water most rapidly?

_____

Reason _____

_____

                                                                       1

(ii) Is water lost from both surfaces of the leaf?

_____

Reason _____

_____

                                                                        1

(b) Name the structures in a leaf through which water is lost?

_____

                                                        1

(c) Name the green pigment found in leaves that convert light energy into chemical energy?

_____

                                                        1

(d) What could be done to improve the reliability of the results?

_____

_____

                                                                1

6. The diagram shows an animal cell.

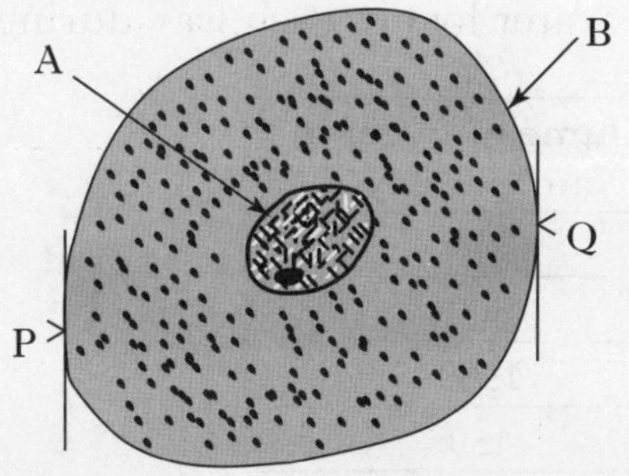

	KU	PS

(a) Name structures A and B.

A _____

B _____   **2**

(b) Name **one** structure present in plant cells that is not present in this animal cell.

_____   **1**

(c) Name one substance that:

(i) Enters the cell by diffusion _____

(ii) Leaves the cell by diffusion _____   **2**

(d) Distance P to Q on the diagram is the diameter of the cell. The distance was measured on three cells using a microscope. The results were as follows:

Cell 1:   63 micrometres

Cell 2:   78 micrometres

Cell 3:   69 micrometres

Calculate the average diameter of these cells.
Space for calculation

_____ *micrometres*   **1**

7. A man loses water from his body in several ways.

The table shows the amount of water lost in each way during a single day.

Method of losing water	Volume of water in cm^3 per day
Breathed out	350
Sweating	500
	1500
Faeces	150

(a) Complete the table by filling in the missing way of losing water.

(b) Draw a bar chart of the information on the grid below.

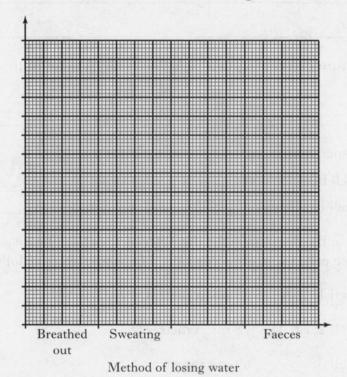

Method of losing water

(c) Name one substance, other than water, that is removed in urine.

_____

(d) Name the structure in the human body that stores urine.

_____

KU	PS
1	
	3
1	
1	

8. Below is a diagram of the human digestive system.

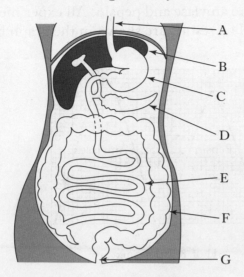

	KU	PS

(a) Why do starch and protein need to be digested?

_____

_____

_____   2

(b) Using letters from the diagram, state where:

   (i)   The soluble products of digestion are absorbed _____

   (ii) Most of the water is absorbed _____   2

9. Experiments were carried out to investigate the action of two enzymes at different pH values. The enzymes were amylase and pepsin. All experiments were carried out at 37°C for 20 minutes. The results are shown in the graph below.

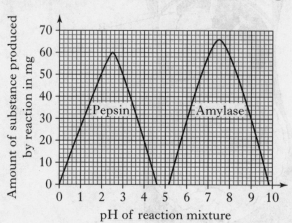

(a) How much substance was produced in the pepsin controlled reaction at pH3?

_____ mg

(b) At which pH values were 60 mg of substance produced by:

(i) pepsin _____

(ii) amylase _____

(c) What chemical substance are all enzymes made from?

_____

(d) If this experiment had been carried out at 60°C, how would the amount of substance produced have compared with these results?

Tick the correct box.

More substance produced

Same amount produced

Less substance produced

KU | PS

1

2

1

1

	KU	PS

10. Evie smokes cigarettes. She has just found out she is pregnant and she finds this information about cigarette smoking and birth weight.

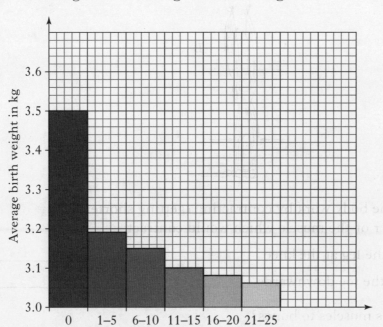

(*a*) Evie smokes 13 cigarettes a day. Use the information in the graph to predict her baby's birth weight.

_____ kg

(*b*) What is the relationship between cigarette smoking and birth weight?

_____

_____

_____

(*c*) What is the difference in average birth weight between a pregnant woman who is a non-smoker and a woman who smokes 7 cigarettes a day?

Space for calculation

_____ kg

(*d*) When a woman smokes 25–29 cigarettes per day the average birth weight is 3.02 kg- add this bar to the graph.

KU/PS marks: (a) 1 (b) 2 (c) 1 (d) 1

11. The diagram shows some of the muscles of a human leg.

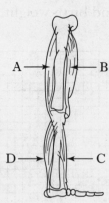

(a) Muscles in the body work by contracting (getting shorter).
Give the letter of the muscle which would contract:

(i) to bend the leg at the knee _____

(ii) to point the toe downwards _____

(b) What attaches muscles to bones?

_____

(c) The hip is an example of a ball and socket joint.
What type of joint is the knee an example of?

_____

KU | PS

2

1

1

12. It is possible to find out how asthma affects breathing by using a peak flow meter. The person blows as hard as possible into the meter as shown in the diagram below.

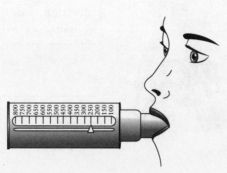

Readings were taken every morning and evening for 7 days from a healthy person and a person with asthma.

The results are shown on the chart below.

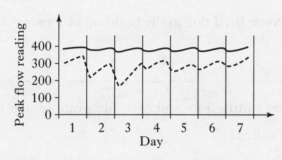

Key ——— Healthy person

----- Person with asthma

(a) On which day was breathing most difficult for the person with asthma?

Day _____

(b) The healthy person's readings are different from the person's with asthma.

State **two** ways in which they are different.

1. _____

2. _____

(c) A person blowing into a peak flow meter obtained a reading of 230.
Does this suggest that this person suffers from asthma?
Give a reason for your answer.

Yes/No _____

Reason _____

_____

KU	PS
	1
	2
	1

	KU	PS

(d) Roy goes to basketball training. During training, many changes take place in his body. These are shown in the diagram.

Pulse rate increases		Breathing rate increases
Breathes more deeply		Produces sweat

(i) Roy's breathing rate increases during the training session. Explain why.

_____

_____   **1**

(ii) Roy's muscles become tired due to the build up of a particular substance. What is it?

_____   **1**

(iii) After training, Roy's pulse rate and breathing rate return to normal.

What is this time called?

_____   **1**

(iv) Choose the correct word to complete the following sentence.
A fit person would be expected to have a long / short recovery time.   **1**

	KU	PS

13. The diagram shows a human eye.

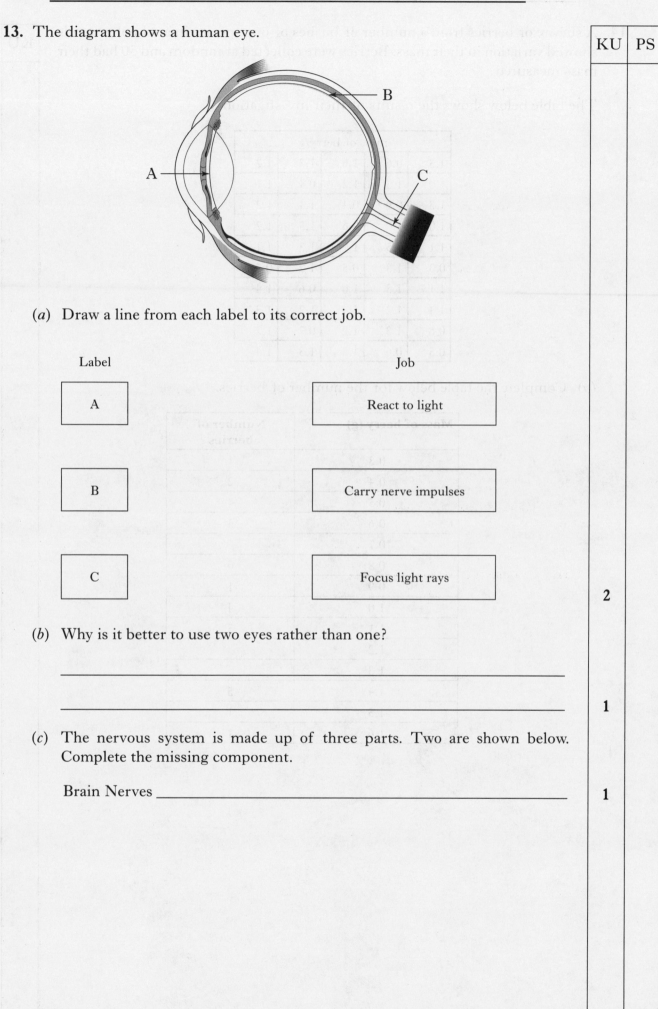

(a) Draw a line from each label to its correct job.

Label

A

B

C

Job

React to light

Carry nerve impulses

Focus light rays

2

(b) Why is it better to use two eyes rather than one?

_____

_____

1

(c) The nervous system is made up of three parts. Two are shown below. Complete the missing component.

Brain Nerves _____

1

14. A survey of berries from a number of bushes of one species in a school grounds showed variation in their mass. Berries were collected at random and 50 had their mass measured.

The table below shows the results of their investigation.

mass of berry/g				
1.3	0.6	1.6	1.3	1.2
1.0	1.3	1.2	0.4	1.1
1.3	0.9	0.4	1.4	1.2
1.0	1.0	0.6	1.5	1.2
1.1	0.5	1.1	1.3	1.1
0.3	1.3	0.5	1.2	0.5
1.1	1.3	1.0	0.6	1.4
1.4	1.2	1.4	1.2	1.3
0.6	1.3	1.2	0.7	1.2
0.5	0.6	1.3	1.3	1.4

(*a*) Complete the table below for the number of berries.

Mass of berry (g)	Number of berries
0.3	1
0.4	2
0.5	4
0.6	5
0.7	
0.8	0
0.9	1
1.0	4
1.1	5
1.2	
1.3	
1.4	5
1.5	1
1.6	1

KU | PS

2

(b) Complete the histogram below using the information.

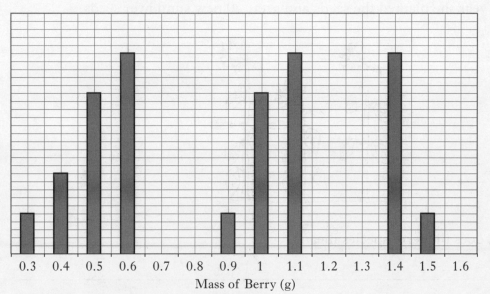

(c) Calculate the ratio of number of berries when the mass is 0.4 g to when the mass was 1 g.

Space for calculation

Answer ————————————— : —————————————

KU PS

3

1

	KU	PS

15. A student collected some sunflower seeds from a flower in her garden. The following year she planted the seeds and grew 20 new sunflower seeds.

(a) What does the term species mean?

_____

_____   1

(b) The student noticed that the plants grew to different heights.
What type of variation is this an example of?

_____   1

(c) She used the tallest plants as parents and collected the seeds from these plants. She hoped to grow even taller plants the following year.

Which of the following terms describes what she is doing?

Tick (√) the correct box.

Cloning	
Genetic engineering	
Selective breeding	
Tissue culture	

1

16. The diagram shows the chromosomes from a body cell of a man.

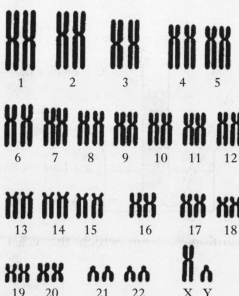

KU | PS

(a) How do you know that these chromosomes must be from a man?

_____

_____

1

(b) How would the chromosomes differ if they had been taken from a sex cell?

_____

_____

1

(c) Name a human condition caused by a chromosome mutation.

_____

1

(d) What procedure can be carried out to detect chromosome characteristics before birth?

_____

1

17. A student investigated the fermentation of solution X by yeast. He used the apparatus below.

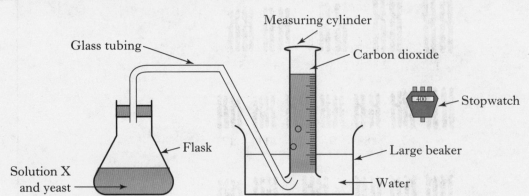

(a) Name the substance in solution X from which the yeast will produce alcohol.

_____

(b) The student investigated fermentation at different temperatures. The results are shown in the table below.

Temperature (°C)	Volume of carbon dioxide collected in 24 hrs (cm³)			Average volume of carbon dioxide collected in 24 hrs (cm³)
10	25	28	28	27
20	58	65	66	63
30	99	108	102	....

(i) **Complete the table** to show the average volume of carbon dioxide collected in 24 hrs at 30°C.

Space for calculation

(ii) Why was the experiment carried out 3 times and an average calculated?

_____

_____

(iii) What do these results tell you about the effect of temperature on the rate of fermentation?

_____

_____

	KU	PS
(a)	1	
(b)(i)		1
(b)(ii)		1
(b)(iii)		1

18. A student investigated the fermentation of different types of sugar by yeast.

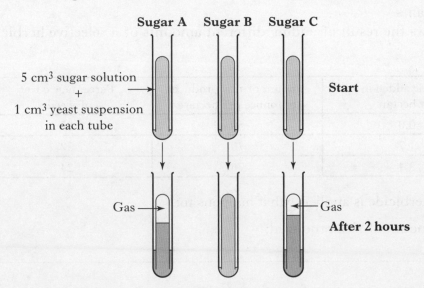

A different type of sugar was put into each tube.
She then placed all 3 sets of apparatus into a water bath at 30°C for 2 hours.
The results are also shown in the diagram.

(a) Give two variables that were kept constant in this investigation.

1. _____

2. _____

(b) Which of the 3 sugars, A, B and C could the yeast ferment?
Give reasons for your answer.

_____

_____

_____

KU | **PS**

2

2

19. A selective herbicide (a type of pesticide) can be used to kill weeds growing among crop plants.

The table shows the result of adding different amounts of a selective herbicide to a rice crop.

Herbicide added in kg per hectare	Amount of rice produced in tonnes per hectare	Percentage cover of weeds
0.0	50	85
1.7	70	32
3.4	76	24

(a) As more herbicide is applied, what happens to:

(i) the amount of rice produced;

_____

(ii) the percentage cover of weeds;

_____

(b) Suggest a reason why rice does not grow well when there are a lot of weeds present.

_____

_____

KU   PS

1

1

1

20. Read the following passage about bird flu and answer the questions that follow.

What is bird flu?

Bird flu was thought only to infect birds until the first human cases were seen in Hong Kong in 1997.

Humans can catch the disease through close contact with live infected birds.

Bird flu produces symptoms which are similar to other types of flu such as fever, sore throats and coughs.

The World Health Organisation said that, by the end of January 2005, there had been 55 confirmed cases of bird flu and 42 deaths in Asia.

There are signs that bird flu can be passed from person to person.

In Thailand a girl who had the disease may have passed the virus to her mother. They both died. The girl's aunt, who was also infected, survived the virus.

Fortunately the normal virus only seems to pass to close relatives and spreads no further.

In Britain, in a typical year between 12000 and 18000 people die from normal flu. This number has been greatly reduced by the flu vaccine, given to 'at risk' people each year. Healthy people should not suffer any complications if they have flu.

The threat of bird flu led to the government ordering supplies of a vaccine. So far, the threat of bird flu has not materialised.

(a) What are the symptoms of bird flu?

_____

_____ | | 1

(b) In Asia, what percentage of confirmed cases died?
Space for calculation.

_____ % | | 1

(c) Why has the number of people who die from flu in Britain fallen?

_____

_____ | | 1

(d) What is an antibiotic?

_____

_____ | 1 |

21. (*a*) Identify the micro-organism used in the production of each of the following products. Draw a line from the product to the correct micro organism.

PRODUCTS	MICRO-ORGANISM
Bread	
Yoghurt	Bacteria
Beer	
Wine	Yeast

KU **2**

(*b*) Decide whether each of the following statements are true or false and if false, give the correction for the wrong word.

STATEMENT	TRUE	FALSE	CORRECTION
Insulin is used to prevent the growth of micro organisms			
Biological detergents contain enzymes.			
Lactic acid is important in the production of bread.			

3

[End of question paper]

Worked Answers

1.

 This question looks at the Biosphere topic and covers food webs

(*a*) (i) Direction of energy flow   **1 mark**

> *HINT*   The arrows in a food chain represent the **direction of energy flow** from one organism to the next. i.e If the caterpillar eats the acorn then the energy from the acorn is transferred to the caterpillar.

(ii) Any of the plants- acorn, beech tree, grass   **1 mark**

> *HINT*   A producer is an organism that makes its own food- all green plants make their own food by photosynthesis.

(iii) Greenfly, caterpillar, worm, shrew, wood mouse, owl, weasel, vole

                                                  1 mark

> *HINT*   A consumer is an organism that eats another organism.

(*b*) (i) Grass- greenfly- shrew- owl
Acorn- caterpillar- wood mouse- owl
Beech tree- worm- vole- weasel-   **1 mark**

> *HINT*   All food chains start with a producer and follow the arrows to the consumer at the top of the food chain.

(ii) Two   **1 mark**
(iii)

Shrew

Ladybird

Greenfly

                                                 1 mark

TOP EXAM TIP

Make sure your food chain makes sense, i.e. don't have a rabbit eating a fox.

2.

 This question covers the Biosphere topic with a question about pollution in a river.

(a) Shrimps   **1 mark**

HINT For each organism, calculate the difference in the value for site A and the value for site B.

Mayfly = 76   Shrimps = 104   Waterlice = 82   Leeches = 94

(b) Domestic or industrial.   **1 mark**

This is a KU question and so you would be expected to be able to recall this.

(c) Tick the second line -   A   low   B   high   **1 mark**

HINT If mayfly larvae like high levels of oxygen then you can assume that low numbers indicate low levels of oxygen, i.e. site A. Vice versa for site B.

(d) 10:1   **1 mark**

This question is on The World of Plants and looks at the uses of plants

3. (a) Raw materials, medicines, food, ornamental.

HINT Raw materials- Plants can be used for building (wood), making clothes (cotton) and other materials such as rubber.

Medicines- Many medicines come from plants such as morphine from poppies.

Food- Plants are a source of food for humans and animals- Examples include wheat, potatoes, carrots.

Ornamental- Many people keep plants in their houses or gardens.

   (b) Chocolate, maize, wheat, potatoes, cabbage, carrots   **1 mark**

   (c) Wood, rubber, cotton   **1 mark**

4. (a) Tropical rainforests and British sugar beet crops.   **1 mark**

   (b) 278 parts per million (ppm).   **1 mark**

   (c) 102 parts per million   **1 mark**

   (d) Maize, rice and soy.   **1 mark**

   (e) Rising temperature and drought.   **1 mark**

   (f) Increased plant growth   **1 mark**

5.

This question relates to the World of Plants topic and mainly covers photosynthesis.

(a) <u>Carbon dioxide</u>   **1 mark**   <u>Oxygen</u>   **1 mark**

(b) Light or Sun   **1 mark**

(c) (i) 5.37

> **HINT** Add all the subjects up and then divide by the number of subjects.

(ii)

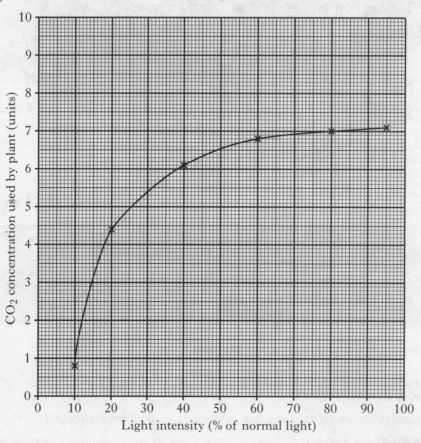

Y axis labelled   **1 mark**
Plots plotted   **1 mark**
Line drawn   **1 mark**

(d) Light intensity   **1 mark**

(e) Temperature   **1 mark**

> **HINT** Type of plant/ size of plant were mentioned in the text so would not be accepted

(f) No.

Reason – would have been better to use more than one plant at each light intensity.

Replicate results increase reliability   **1 mark**

> **HINT** Anything that suggests replication would be accepted.
> Reason is the important part – just writing Yes or No would give no marks.

(g) Use more light intensity values   **1 mark**

This question covers the Animal Survival topic – the Need for Food.

The digestive system and its parts should be known.

6. (a) B

Liver

Large intestine

H   4 correct = **2 marks**

3 or 2 correct = **1 mark**

1 correct = **0 marks**

(b) F   **1 mark**

HINT   The answer large intestine will be accepted- but not intestine.

(c) Any 2 of:

Long / large surface area (villi)   / good blood supply

1 mark for each = **2 marks**

HINT   Large surface area and villi would count as 1 answer so do not write both of these.

(d) (i)

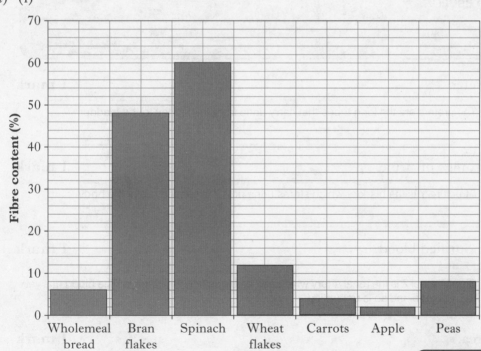

Draw missing labels on the x axis   **1 mark**

Draw missing bars   **1 mark**

TOP EXAM TIP

On a bar graph make sure that you draw an obvious line across the top of the bar.

You **will** be penalised if the top of the bar is not obvious.

(ii) 20 **1 mark**

> **HINT** To calculate the average add up all the numbers and divide by 7.

(iii) 4 : 1 **1 mark**

> **HINT** 48: 12 can be simplified by dividing each number by 12.
>
> Make sure you put the answer you want to give in the space provided to you.

(iv) 15% **1 mark**

> **HINT** 60% = 60g in 100g of spinach.
>
> So in a 25g portion it would be
>
> 100/25 x 60 = 15g

> **HINT** Calculations like this are common.
>
> Make sure you use the correct value from the table (underline words in the question to help)
>
> Once you've done the calculation – think about whether the answer makes sense – for example in this question if the answer you get is greater than 25 – it can't possibly be right!

7.

 This question is on Animal Survival and looks at the Water and Waste section.

(a) (i) Ureter. **1 mark**

> **HINT** The ureters are the long tubes that carry urine from the kidney to the bladder.

(ii) Strorage of urine. **1 mark**

> **HINT** Structure Y is the bladder which stores urine until it is passed out of the body.

(iii) Unpurified blood. **1 mark**

> **HINT** Blood vessel Z is the renal artery which carries blood into the kidneys. The blood has yet to be filtered and so is unpurified.

(iv) Urea. **1 mark**

> **HINT** Urea is removed from the blood by the kidneys along with other waste. However, urea is a toxic waste and if it is allowed to build up in the body it can cause death. It is made in the liver by the breakdown of amino acids. When urea combines with water it makes urine.

(b) (i) Breath= 300 cm³

Respiration= 400 cm³ **1 mark**

> **HINT**
>
> | | |
> |---|---|
> | If you add together- sweat (600), urine (1200) faeces (300) it gives you a total of =2100. The total needed is 2400 so 2400-2100=300. So breath=300. | Respiration=400 If you add together- Food (900), Drink (1100) =2000 The total needed is 2400 so 2400-2000=400. So respiration=400. |

(ii) Water gain is equal to water loss. **1 mark**

> **HINT** Both water gain and water loss are 2400 cm³, this shows that water balance has been achieved.

(iii) 2:1. **1 mark**

> **HINT**
>
> Water lost as urine is 1200 cm³, water lost as sweat is 600 cm³.
>
> To get these numbers as ratios we must simplify them down as much as possible. So what is the largest number both of these can be divided by? 600.
>
> 1200 divided by 600 = 2.
>
> 600 divided by 600 = 1.
>
> So the ratio 2:1 is given.

(iv) 25%

> **HINT** 600 divided by 2400 times 100.

This question refers to Investigating Cells and is looking at the structure of a cell.

8. (a) Animal cell. **1 mark**

> **HINT** What structures are visible? - Cell membrane, cytoplasm and nucleus, both animal and plant cells have these but plant cells also have a variety of other structures such as chloroplast and a cell wall. So this is an animal cell.

(b) Animal cell- Cell membrane, cytoplasm and nucleus. **1 mark**

Plant cell- Cell membrane, cytoplasm, nucleus, chloroplast, cell wall. **1 mark**

> **HINT** Only plant cells have a cell wall and chloroplasts.

TOP EXAM TIP

Learn the structure of plant and animal cells and their functions.

9.

This question refers to diffusion in the Cells topic.

(a) Diffusion.   **1 mark**

(b) Membrane   **1 mark**

(c) Glucose and oxygen   **2 marks**

(d) Osmosis   **1 mark**

HINT — For general level you only need to know that osmosis is the diffusion of water. You don't need to know the details

10.

This question covers the Investigating Cells topic and is primarily about the properties of enzymes.

(a) 20 minutes   **1 mark**

(b) 40   **1 mark**

HINT — Both these questions involve reading values from the graph so make sure you read the question carefully.

(c) The enzyme was denatured.   **1 mark**

HINT — A reference to the fact that enzymes do not work at high temperatures would be enough. Do **not** write that the enzyme is killed.

(d) 0–10°C   **1 mark**

11.

This question covers the Body in Action topic. The Movement sub-topic is also covered.

(a) (i) D   **1 mark**
    (ii) B   **1 mark**
    (iii) A   **1 mark**

(b) Movement / protection   either answer   **1 mark**

(c) prevent friction at end of bone.   **1 mark**

HINT — For this answer it is sufficient to describe the function of parts X and Y without naming them. Do not just write the name of the part as you are not being asked for that.

12.

This question is primarily a problem solving question, taking information from a graph. There is a KU question at the end.

(a) Number of people needing transplants is higher than the number of dead donors   **1 mark**

refer to a value from the graph   **1 mark**

or

Number of people needing transplants is higher than the number of transplants carried out at the moment.   **1 mark**

refer to a value from the graph   **1 mark**

(b) Live donors are used.   **1 mark**

(c) There is less trauma than getting an operation.   **1 mark**

13.

This question looks at The Body In Action and covers The Need For Energy topic

(a) Vessel D   **1 mark**

> **HINT** Vessel D is the aorta, the main artery leaving the heart carrying blood to the rest of the body- REMEMBER A AND A- The Aorta is the main Artery.

(b) F and G.   **1 mark**

> **HINT** Ventricles are the two lower chambers of the heart. The blood flows into the atria and then down into the ventricles.

(c) Vessel B.   **1 mark**

> **HINT** Vessel B is the pulmonary artery, this takes blood from the right ventricle to the lungs where it becomes oxygenated.

(d) Stop blood flowing backwards.   **1 mark**

(e) It has to pump blood all around the body.   **1 mark**

> **HINT** The right side just has to pump blood to the lungs.

14.

This question looks at The Body In Action and is a problem solving question on the Changing Levels of Performance section

(a) (i) Label- Pulse rate (beats per minute). **1 mark**

(ii) Point plotted correctly. **1 mark**

(b) At 8 minutes both the untrained and trained athletes' pulse rates have not returned to the resting pulse rate. **1 mark**

HINT As the pulse rates are above the resting rate that the athletes started at, this shows the athletes had not yet fully recovered from the exercise.

(c) Started lower, did not increase as much during exercise. **1 mark**

HINT The trained athlete had a lower starting pulse rate than the untrained athlete, this did not increase as much as the untrained athletes did during exercise. This shows the difference in fitness between the two athletes.

(d) Continue timing until the pulse rate returns to 65 (starting pulse rate before exercise). **1 mark**

HINT This will give an accurate measurement of recovery time, as the full time during exercise and time taken to recover is measured.

15.

This question is a problem solving question and is linked to the body in Action topic and the Need for Energy in particular.

(a) Rose and then fell **1 mark**

Mention values rose to 55 per 100 000

Or

Fell to 17 per 100 000 **1 mark**

Other appropriate values would be accepted.

(b) 14 per 100 000 **1 mark**

(c) 19 per 100 000 **1 mark**

1980   –   Men = 35 per 100 000

Women = 16 per 100 000

35 – 16 = 19

16.

This question is on Inheritance and covers the What is Inheritance topic

(a) (i) F1. **1 mark**

> HINT First generation is written as F1.

(ii) All round. **1 mark**

> HINT As round is the dominant phenotype and both the parents are true breeding, all of the first generation seeds will be round in appearance.

(b) STATEMENT 1- False, PHENOTYPE. **1 mark**

STATEMENT 2- True. **1 mark**

STATEMENT 3- True. **1 mark**

> HINT The phenotype is the physical appearance of an organism.
>
> REMEMBER P AND P- Phenotype and Physical appearance
>
> REMEMBER G AND G- Genotype and Genetic makeup.

(c) (i)

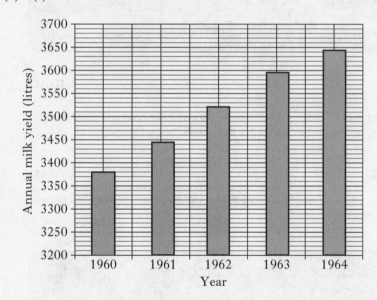

Label- Annual milk yield (litres). **1 mark**

Scale correctly added **1 mark**

Bars added correctly **1 mark**

TOP EXAM TIP

When doing bar graphs, make sure an even scale is added that goes up in even numbers- (0,2,4,6,8)

Make sure your graph starts at 0.

Make sure your bars are spaced apart by a least one box between each of them and the tops of all bars are marked clearly.

(ii) The milk yield increases from 1960–1965. **1 mark**

(iii) 3517. **1 mark**

TOP EXAM TIP

DESCRIBE QUESTION- When asked to describe a trend make sure you describe what is happening to the numbers.

 HINT

To get an average add up all of the numbers and divide by the number of subjects so-

3379 + 3446 + 3521 + 3596 + 3643 = 17585

17585 divided by 5 = 3517.

17.

This question refers to Biotechnology.

The uses of different kinds of micro- organisms and the safe handling of microbes will be covered.

(*a*) Yeast

Bacteria **1 mark**

(*b*) B   D   E   A   C **2 marks**

18.

This question is on Biotechnology and refers to genetic engineering.

(*a*) (i) X is a gene. **1 mark**

 HINT

The circular chromosome of a bacterium is similar to ours as it is also made up of genes. The small units making up the chromosome are the genes.

(ii) Genetic engineering. **1 mark**

 HINT

Genetic engineering is used by scientists to transfer genetic material such as genes from one organism to another.

This question is on Biotechnology and is a problem solving question based on the yeast section

(b) (i) Temperature. **1 mark**

> **HINT** As the same experiment was carried out at six different temperatures we can see that the variable that has been changed is temperature.

(ii) Six. **1 mark**

> **HINT** Six different temperatures were used- 10, 20, 30, 40, 50, 60 °C.

(iii) Volume of glucose solution, size of measuring cylinder. **1 mark**

> **HINT** Only one variable can be changed in each experiment in order to keep the results reliable.

(iv) Repeat the experiment and take an average of the results, use more subjects. **1 mark**

WORKED ANSWERS   PAPER B

1.

> This question looks at the Biosphere and is based on food webs.

(a) Cockles are eaten by mussels and walruses- TRUE

Animal plankton is not eaten by anything- FALSE

Seals eat cod and mussels- TRUE

All correct= **1 mark**

(b) PRODUCER- An organism that makes its own food **1 mark**

> **HINT** A producer is an organism that produces its own food by photosynthesis and is always a plant.

CONSUMER- An organism that feeds on other organisms **1 mark**

(c) ANIMAL PLANKTON, COCKLES, WALRUS, SEAL, HERRING, COD **1 mark**

(d) TWO **1 mark**

> **HINT** Plant plankton- animal plankton- cockle- walrus
>
> Plant plankton- animal plankton- cockle- mussel- seal- walrus.

(e) Movement, faeces/waste, heat, any two   **1 mark**

HINT The only form of energy that is transferred along a food chain is energy used for growth.

(f) The arrows represent the direction of energy flow   **1 mark**

2.

This is a Biosphere problem solving question looking at the effect of light intensity on the growth of a plant.

(a)

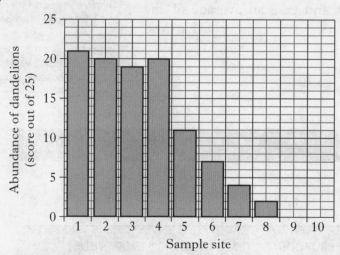

Y axis scale added   **1 mark**

Y axis labelled   **1 mark**

Line drawn   **1 mark**

TOP EXAM TIP

Make sure an even scale is added that goes up in even numbers- (0,2,4,6,8) and it is labelled.

Make sure your graph scale starts at **0** but do not extend your line to **0** unless there is a point there.

Plot all points correctly

For all graphs think- **LABELS, SCALES, POINTS.**

(b) (i) 10.4   **1 mark**

HINT 21 + 20 + 19 + 20 + 11 + 7 + 4 + 2 = 104.

Divide 104 by 10 = 10.4.

TOP EXAM TIP

When asked to describe a relationship there is no need to explain the reason behind it. This is normally answered as- as the.....increases, the ...... decreases (depending on the results).

(ii) As light intensity decreases, the abundance of dandelions also decreases   **1 mark**

(iii) Temperature, moisture content of the soil, pH of the soil   **1 mark**

HINT Abiotic factors are non-living factors such as light and temperature.

Biotic factors are living factors such as predation and competition.

3.

This is a question looking at keys in the Biosphere topic.

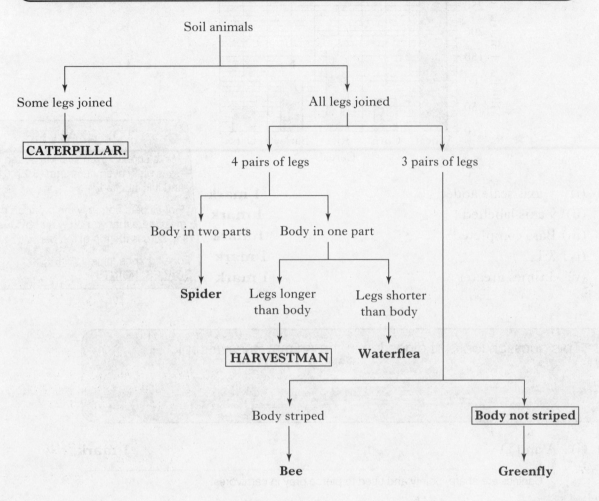

3 mark

(b) Stripes/ one has stripes/ one does not have stripes **1 mark**

(c) All legs joined AND 4 pairs of legs AND body in 1 part **1 mark**

4.

This question looks at the structure of a plant from the World of Plants topic.

(a) (i) A **1 mark**

    (ii) NAME- Stigma

         FUNCTION- Region to which pollen grains attach **1 mark**

    (iii) E **1 mark**

    (iv) Seed **1 mark**

         Fruit **1 mark**

    (v) Oats **1 mark**

> **HINT** In a wind pollinated plant, the structures hang out so that pollen can be blown away from the anthers of one plant and be caught on the stigma of other plants. An insect pollinated plant has its structures securely inside.

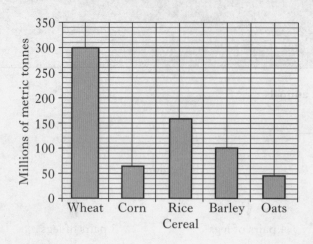

TOP EXAM TIP

Make sure an even scale is added that goes up in even numbers- (0,2,4,6,8) and it is labelled.

Plot all bars correctly, on a bar graph make sure that you draw an obvious line across the top of the bar.

For all graphs think- **LABELS, SCALES, POINTS**.

(b) (i) y axis scale added   **1 mark**

     (ii) y axis labelled   **1 mark**

     (iii) Bars completed   **1 mark**

     (iv) 3:1   **1 mark**

     (v) 3 times greater   **1 mark**

5.

 This question looks at digestion in the Animal Survival topic.

(a) (i) A and D   **1 mark**

> **HINT** Canines are sharp, pointy and used to pierce prey in carnivores.
>
> They are only found on the lower jaw in herbivores.

     (ii) F   **1 mark**

> **HINT** The molars are used in herbivores to grind plant material.

     (iii) A   **1 mark**

> **HINT** Canines are used to pierce prey in carnivores.

(b) (i) Salivary glands- P

          Stomach- R

          Large Intestine- T

          Liver - Y   All correct   **2 marks**

     (ii) Insoluble

          Soluble   Both correct   **2 marks**

> **HINT** Digestion breaks down large particles into small particles and this allows food to be absorbed into the bloodstream.

    (iii) The large intestine absorbs water **1 mark**

6.

 This question examines enzymes in the Investigating Cells topic.

> **TOP EXAM TIP**
>
> Remember the pneumonic- SAM for **S**tarch, **A**mylase, **M**altose. Make up one for each enzyme equation and other equations you need to know, this will help you to remember them.

(a) (i)   large   small   –   **1 mark**

    (ii) A catalyst speeds up chemical reactions **1 mark**

    (iii) Protein   **1 mark**

    (iv) Pepsin/lipase   **1 mark**

    (v) The colour of the Benedict's solution is orange when mixed with the liquid showing sugar is present
        1 mark

> **TOP EXAM TIP**
>
> Learn the three common body enzymes, where they work and each of their reactions.

(b) (i) The enzyme activity increases as the temperature increases **1 mark**

At about 40°C the activity of the enzyme starts to decrease rapidly **1 mark**

> **HINT** The best temperature for activity of enzymes is at 37° C, which is body temperature.

    (ii) 30–40°C   **1 mark**

    (iii) 20 units   **1 mark**

7.

 This is a problem solving question based on Responding to the Environment.

(a) 28   **1 mark**

> **HINT** 24 + 30 + 30 = 84. 84/3 = 28.

(b) DECREASES   **1 mark**

    DECREASES   **1 mark**

> **HINT** Look for a pattern in the results.

> **TOP EXAM TIP**
>
> If you are asked how to improve the reliability of the experiment- always say repeat the experiment.

(c) Move less/ stop moving.   **1 mark**

(d) Repeat the experiment.   **1 mark**

8.

This question looks at mitosis in the Investigating Cells unit.

(a) (i) C, A, D, B   **1 mark**
   (ii) Nucleus   **1 mark**

     Increases   **1 mark**

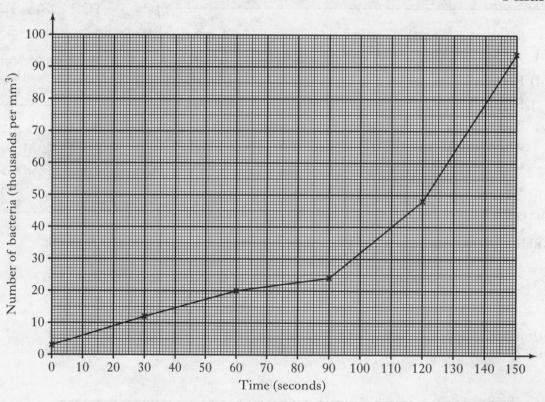

(b) (i) y axis scale added   **1 mark**
      y axis labelled   **1 mark**
      Line drawn   **1 mark**

   (ii) 1:2   **1 mark**

   (iii) 32 + /-1 thousands per mm^3

> **TOP EXAM TIP**
>
> Make sure an even scale is added that goes up in even numbers- (0,2,4,6,8) and it is labelled.
>
> Make sure your graph scale starts at **0** but do not extend your line to **0** unless there is a point there.
>
> Plot all points correctly
>
> For all graphs think- **LABELS, SCALES, POINTS**.

9.

The parts of the eye and their functions from the Body In Action topic are covered in this question.

> **TOP EXAM TIP**
>
> When asked to do a ratio, try to divide both numbers by the lowest number that they can both be divided by.

(a)

LETTER	NAME	FUNCTION
A	Cornea	Allows light to enter and focuses it.
E	Optic nerve	Carries nerve impulses from retina to brain.
D	Retina	Light sensitive layer.
C	Lens	Flexible structure, focussing light onto retina.
B	Iris	Coloured part of the eye.

All Correct   **3 marks**

(b) As both eyes   **1 mark**

10.

This question looks at parts of the lungs and their functions in the Body In Action topic.

(a)

LETTER	NAME	FUNCTION
C	Alveoli	Air sacs allowing oxygen to pass to the lungs.
B	Bronchi	Two divisions of windpipe.
A	Trachea	Tube from mouth to bronchi

All correct   **3 marks**

(b) Rib cage- Heart

Vertebrae- Spinal cord

Skull- Brain   All correct-   **2 marks**

11.

This is a problem solving question based on pollution.

(a) Agricultural   **1 mark**

(b) 12   **1 mark**

(c) 6   **1 mark**

HINT ▷   10-4 = 6.

(d) Complete the bar graph correctly   **2 marks**

12.

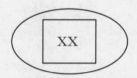 This question looks at sex chromosomes and inherited characteristics in the Inheritance topic.

(*a*) (i)

MALE

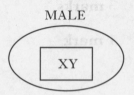

1 mark for each correct answer- **2 marks for all answers correct.**

 HINT > Females gamete mother cells have the genotype- XX So can only pass on X as a gamete. Male gamete mother cells have the genotype XY So can pass on X or Y as a gamete.

(ii)	Two	**1 mark**
(iii)	Gametes	**1 mark**
(iv)	Fertilisation	**1 mark**
(v)	Phenotype	**1 mark**
(*b*)	Species	**1 mark**
	fertile	**1 mark**

13.

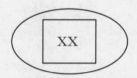 This is a problem solving question based on Inheritance.

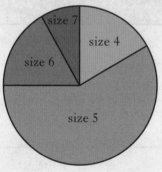

School pupils' shoe sizes

(*a*) All divisions correctly marked   **2 marks**

(b) Discontinous variation   **1 mark**

> HINT Discontinuous variation is when the factor examined can be spilt up into two or more distinct groups e.g Blood groups.
>
> Continuous variation is when the factor varies from one extreme to another e.g. height.

(c) 58%   **1 mark**

> HINT 70 people divided by 120 gives you 0.58
>
> Times 0.58 by 100 = 58%.

14.

> ★ This is a question looking at Sewage treatment in the Biotechnology topic.

(a) (i) Methane   **1 mark**

> HINT Methane is made at sewage plants and can be reused as a gas for a variety of uses such as cooking and heating.

(ii) They do not cause damage to the environment like fossil fuels do
                                                                     1 mark

(b) (i) A   **1 mark**

> HINT The bacterial count is very high here and the oxygen concentration is low as the high number of bacteria are using it up, this shows that the sewage has been put in here.

(ii) The bacteria are high and the oxygen concentration is low   **1 mark**
(iii) High bacteria concentration, low oxygen concentration   **1 mark**
(iv) When the oxygen concentration is high the number of fish is high.

15.

> ★ This question looks at sterile techniques in the Biotechnology topic.

STERILE TECHNIQUE	REASON
Sterile Petri dishes kept closed.	Prevent entry of contaminants
Wire loop flamed before use.	To destroy unwanted microbes
Work surface cleaned with disinfectant.	To kill microbes
Hands washed.	Remove micro-organisms
Lab coat worn.	Protect clothing

5 KU- 1 mark for each correct answer.

16.

This question is a problem solving question looking at information on Antibiotics from a passage.

(a) Antibiotics   1 mark

(b) Penicillium   1 mark

(c) Penicillin   1 mark

(d) Pneumonia/chest infections   1 mark

(e) Viruses   1 mark

WORKED ANSWERS   **PAPER C**

1.

This question is on the Investigating an Ecosystem part of the Biosphere.

(a) A- Habitat   1 mark

(b) B- Producer   1 mark

(c) A- Habitat and D-Community   1 mark for both

(d) H- Light intensity   1 mark

2.

This question covers the Biosphere (How it works) and is about Food Webs

(a) Heron   / fish   / diving beetle / frog

     Fish   / fly larva/water flea / water flea   / water boatman

                                                                                 1 mark

HINT The predator and prey must relate to each other, ie the predator is above the prey in the answer

(b) Frog   1 mark

(c) Algae / duckweed   1 mark

> **HINT** Producers are ALWAYS plants

(d) Two **1 mark**

> **HINT**
> Algae ⟶ Fly larva ⟶ Fish ⟶ Heron
> Duckweed ⟶ Fly larva ⟶ Fish ⟶ Heron

⭐ This question is a Problem solving question that involves predators and prey

TOP EXAM TIP

No value for 0 has been given so DO NOT join point to 0 in the graph, you will lose a mark for this.

3.

(a) Points plotted correctly **1 mark**
   Line drawn **1 mark**

(b) 24000 thousand **1 mark**

> **HINT** This value is obtained by reading the graph for snowshoe hare and finding the value on the y-axis for 1921.

(c) 30 **1 mark**

> **HINT** After drawing your line for lynx, you need to predict what the value in 1925 would have been.

(d) (i) Fall **1 mark**
    (ii) It is following the snowshoe hare pattern **1 mark**

> **HINT** This question requires you to look for and find a pattern between the 2 animals

4.

⭐ This question covers Keys part of the Biosphere

TOP EXAM TIP

Key questions tend to be done poorly, make sure you check that your answers make sense after doing them.

(a) Grey body   green head **2 marks**

(b) Food / space **1 mark**

5.

This question is a problem solving question about Pollution

(a) All parts given correct size **2 marks**

Each part labelled correctly **1 mark**

(b) 45 **1 mark**

> **HINT**
> Calculation   From the table 15% caused by dust
> So 15 % of 300 = 45

(c) 3 **1 mark**

> **HINT**
> Calculation   45 in 1980   15 in 2000
> So 45/15 = 3

6.

This question covers Animal Survival (the need for Reproduction)

TOP EXAM TIP

If the question states 'From the diagram.....' then an answer, even if it is correct, that is not obvious from the diagram will NOT be accepted

(a) (i) Has a tail **1 mark**

(ii) Allows it to move **1 mark**

(b) There is only a small chance of reaching egg / to maximise chances of reaching egg **1 mark**

(c) 6 **1 mark**

> **HINT**
> Sex cells have half the number of chromosomes of a body cell

(d) The female sex cells are made in the OVARY and are called OVULES **1 mark**

The male sex cells are made in the TESTES and are called SPERM **1 mark**

(e)

LETTER	NAME	FUNCTION
B	OVARY	Where eggs are made
C	Uterus	Where the fetus would develop.
E	TESTES	Where sperm are made
G	Penis	MALE SEX ORGAN

4 MARKS- EACH CORRECT ANSWER **4 marks**

7.

    (a) Light   **1 mark**

    (b) The woodlice move away from the light side into the dark side   **1 mark**

> **HINT** You are being asked to describe what direction the woodlice are moving in from the information given.

    (c) To improve the reliability of the results   **1 mark**

> **TOP EXAM TIP**
>
> This a very common exam question- When asked why large numbers of subjects are used in an experiment, why the experiment is repeated or why an average is taken of the results, the reason is always-
>
> TO IMPROVE THE RELIABLITY OF THE RESULTS.

    (d) So the woodlice have time to settle/ to adjust to conditions/ explore conditions   **1 mark**

    (e) Response of maggots to light/ paramecium to acid   **1 mark**

> **TOP EXAM TIP**
>
> Another very common exam question. The answer is always- to allow time for the subject to adjust/ experiment to work.

8.

> This question is a problem solving question based on reading information from a passage. This is a very common exam question.

> **TOP EXAM TIP**
>
> Learn at least two animal responses to environmental conditions as this is a common question.

    (a) Over the Himalayan Mountains   **1 mark**

    (b) Any two of- Atlantic, Mississippi, Central and Pacific flyways   **1 mark**

    (c) Geese AND waterfowl   **1 mark**

    (d) Arctic Tundra   **1 mark**

    (e) 5,000 miles   **1 mark**

    (f) White with jet black wing tips   **1 mark**

9.

> This question is a Problem solving question.

    (a) (i) Increased   **1 mark**
          (ii) England   **1 mark**

    (b) 9 %   **1 mark**

> **HINT**
> Calculation   1978 – 170 otters found
>
> 1888 total number

So $170/1888 \times 100 = 9\%$ **1 mark**

(c) 29%   **1 mark**

> **HINT**
> Calculation   1985 = 1717   1993 = 2211
>
> Increase = 2211 – 1717 = 494
>
> % increase: 494/1717 x 100 = 29%

TOP EXAM TIP

These are often answered badly.

Remember –   1.Calculate how many
it has gone up by

2. Divide this by the ORIGINAL

3. Multiply by 100

(d) 322   **1 mark**

> **HINT**
> 529 – 207 = 322

(e) Increase   **1 mark**

10.

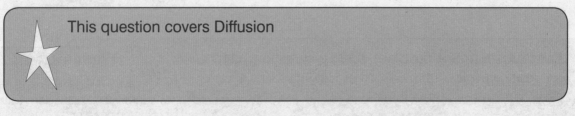

This question covers Diffusion

(a)   mouth/nose

Carbon dioxide

Small intestine   All
correct-   **2 marks**

(b) B

(c) Food/glucose   **1 mark**

(d) Membrane   **1 mark**

(e) Movement of water molecules from area of High Water Concentration to area of Low Water Concentration across a semi permeable membrane

1 mark

11.

This question looks at Aerobic Respiration in the Cells topic

(a) Any two of- Growth, movement, cell division, uptake of chemicals

2 marks

HINT > Learn at least two reasons why cells need energy as this is a common exam question.

(b) The type of food used **1 mark**

HINT > The variables are the parameters investigated in the experiment. In this case, there are three different types of food used so the variable being changed is type of food.

(c) Any two of- Volume of water, distance of food being burned from the bottom of the test tube, size of test tube **1 mark**

TOP EXAM TIP

This is a very common exam question, all other variables other than the one being investigated must be kept the same in order for the results of the experiment to be reliable.

(d) 8.4 kJ **1 mark**

HINT > 4.2 kJ is needed to raise the water by 1°C so in order to raise the water by 2°C double this is needed- 8.4 kJ.

(e) Oxygen **1 mark**

HINT > Oxygen is required by cells in order to release the energy contained in foods.

12.

 This question looks at the structure of a broad bean seed. Growing Plants section of the Plants topic.

(a) A- EMBRYO **1 mark**

B- SEED COAT **1 mark**

(b) C- FOOD STORE **1 mark**

(c) Protects the plant **1 mark**

TOP EXAM TIP

The structure of a seed and structure of a plant are common questions- so make sure you learn the different parts and functions.

13.

(a) X- Stamen **1 mark**

Y- Stigma **1 mark**

(b) Catches pollen **1 mark**

(c) Attracts insects **1 mark**

14.

This question covers Body in Action (Coordination) and is about the ear

TOP EXAM TIP

Both the structures of the ear and of the eye are common exam questions- make sure you learn the parts and functions of both.

(a) B

    Cochlea

           Balance

    Auditory nerve   sends impulses to the brain   **3 marks**

(b) It is easier to judge direction of sound   **1 mark**

15.

This question is a Problem Solving question and covers inheritance (Variation)

(a) Teeth worn   skull bones fused   both needed   **1 mark**

(b) She cooked food   **1 mark**

(c) Isolation   poor resources   few predators   **1 mark**

(d) The homo erectus population   **1 mark**

(e) 18000 years old.   **1 mark**

16.

This question looks at the different parts of blood in the Need For Energy part of Body In Action.

(a) COMPONENT   DESCRIPTION

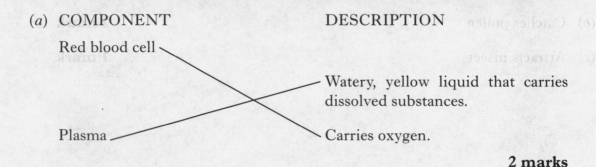

Red blood cell

Plasma

Watery, yellow liquid that carries dissolved substances.

Carries oxygen.

    2 marks

(b)

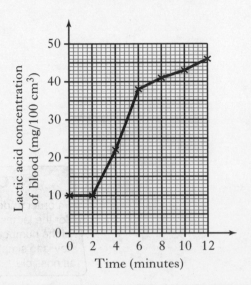

(i) Y axis scale added   **1 mark**

(ii) Y axis labelled   **1 mark**

(iii) Line drawn   **1 mark**

(iv) As time increases so does the lactic acid concentration   **1 mark**

> **TOP EXAM TIP**
>
> Make sure an even scale is added that goes up in even numbers- (0,2,4,6,8) and it is labelled.
>
> Make sure your graph scale starts at **0** but do not extend your line to **0** unless there is a point there.
>
> Plot all points correctly
>
> For all graphs think- **LABELS, SCALES, POINTS.**

> **HINT** This is a describe question, you are being asked to describe the relationship between time and the lactic acid concentration of the athlete, you do not have to explain why this occurs.

> **TOP EXAM TIP**
>
> With describe questions you only have to state the trend that is shown, if you are asked to explain it then you have to give a reason for the trend.

(v) 30   **1 mark**

> **HINT** Add up all of numbers and divide by the number of entries.

17.

 This question is looking at Inheritance topic

(a)

TERM	DEFINITION
Phenotype	Always expressed if present
Genotype	Physical appearance
Dominant	Genetic makeup of an organism

2 marks

(b) (i) 16%   **1 mark**

     (ii) 4B:1AB   **1 mark**

> **HINT** 4-16 divided by 4=4 and 4 divided by 4 is 1 giving the ratio 4:1.

(iii)

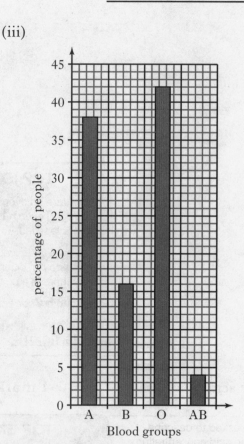

percentage of people

Blood groups

1 mark

18.

 This question looks at the Detergents section of the Biotechnology topic.

(a) Biological **1 mark**

HINT › The best detergent will be the one that has been the most effective at removing the stain, in this case the biological one.

(b) The type of washing powder **1 mark**

HINT › As both biological and non-biological washing powders were used in the experiments, the variable changed is the type of washing powders.

(c) Any two of- volume of water/mass of detergent/size of cloth/time

1 mark

(d) Biological washing powders contain enzymes and non-biological powders do not **1 mark**

19.

> This question looks at the Reprogramming Microbes section of Biotechnology.

(*a*) Genetic engineering   **1 mark**

(*b*) PRODUCT- Insulin   **1 mark**

   USE- Treat diabetes   **1 mark**

(*c*) Asexual   **1 mark**

(*d*) Cheese/yoghurt   **1 mark**

WORKED ANSWERS   PAPER D

1.

> This question covers the biosphere and is a Problem solving question with one question on KU.

(*a*) 100000   **1 mark**

> *HINT*   Calculation: if there are an average of 20 in each 1 m² quadrat then to work out for the whole wood you would multiply
>
> 20 x 5000 = 100000 bluebells.

(*b*) Student   Mike

   Reason   He threw more quadrats.   **1 mark**

   (both answers needed to get the mark)

> *HINT*   The more often you throw the quadrat, the more reliable your results will be.

TOP EXAM TIP

The question of repeating experiments to improve reliability comes up every year in some shape or form.

Remember REPEATING MEANS MORE RELIABLE

(*c*) Pitfall trap

   A pitfall trap is used to catch crawling insects   **1 mark**

(*d*) Temperature, moisture, pH   **1 mark**

2.

This question covers The biosphere and is about Food Webs. These are very common in the exam

(*a*) (i) Consumer – locust/impala/tick/seed eating bird/tick bird/snake/hawk/ scorpion/baboon/leopard **1 mark**

> *HINT*   Consumers are all organisms in a food web that feed on another organism

    (ii) Producer – grass **1 mark**

> *HINT*   Producers are always plants as they can make their own food by photosynthesis

(*b*)   grass   locust   scorpion   baboon **1 mark**

> *HINT*   Make sure your food chain contains the right number of organisms. Each space in a question like this needs to be completed.

TOP EXAM TIP

Do NOT write that the arrows show what is eaten by what – this is a very common mistake.

All food chains should begin with a plant

(*c*) The direction of energy flow. **1 mark**

3.

This question covers a few parts of the course – mainly the World of Plants (Growing Plants) but there is also a question about Investigating Cells (Diffusion). It covers both KU and PS

(*a*) To contain the sulphur dioxide when it was produced/prevent entry of any other gases **1 mark**

(*b*) Temperature / dampness of cotton wool/ size of dish **2 marks**

There will be other answers that would be accepted as long as they were reasonable.

(*c*) Diffusion **1 mark**

> *HINT*   Diffusion is the movement of molecules from an area of high concentration to an area of low concentration

(d) (i) 75%   **1 mark**

TOP EXAM TIP

Percentages come up quite often.
Think about the numbers you are
using and when you have an answer
think about whether it is reasonable!
e.g. if 15 out of 20 germinated your
answer must be more than 50%!

(ii) Reduced germination   **1 mark**

HINT   It would not be right to say that germination was stopped as
some seeds did germinate. However, less did germinate.

(e) 1. Water   **2 marks**

2. Warmth

HINT   For seeds to germinate remember WOW

Water   Oxygen   Warmth

4.

This question covers The World of Plants (Making food). Information
has to be extracted from a graph and general photosynthesis questions
are asked.

(a) 15.00   **1 mark**

HINT   This answer is just taken straight from the graph – where the line is at its highest.

(b) As starch   **1 mark**

HINT   The glucose produced during photosynthesis can be stored as starch.

(c) 0900 – 1200   **1 mark**

HINT   The part of the graph where it is rising steepest is the time period required

(d) Phloem   **1 mark**

HINT   Water travels in xylem and food travels in phloem

5.

(a) (i)   lower

Reason   when the lower surface is covered in jelly,
more water is kept in making the leaf heavier.   **1 mark**

HINT | You need to compare Leaves B and C for this answer. When jelly is spread on the surface it keeps the water in, so the lower surface must lose more water (leaf B is heavier than leaf C)

    (ii) yes

         Reason   all leaves lost water   **1 mark**

(b) Stoma   **1 mark**

(c) Chlorophyll   **1 mark**

(d) Repeat   **1 mark**

6.

 This question covers Investigating cells (Investigating living cells)

(a) A - Nucleus

      B - Membrane   **2 marks**

(b) Cell wall / vacuole/ chloroplast   **1 mark**

HINT | Plant cells and animal cells both have a nucleus, cytoplasm and membrane. Plant cells also have a vacuole, a cell wall and chloroplast.

(c) (i)   Food/glucose/oxygen
      (ii) Carbon dioxide/waste   **2 marks**

(d) 70 micrometres   **1 mark**

HINT | Calculation – to work out an average add up all the numbers and divide by how many numbers you have.

63 + 78 + 69 = 210

210/3 = 70

7.

 This question covers the Animal Survival topic (the Need for Water). It requires a graph to be drawn

(a) Urine   **1 mark**

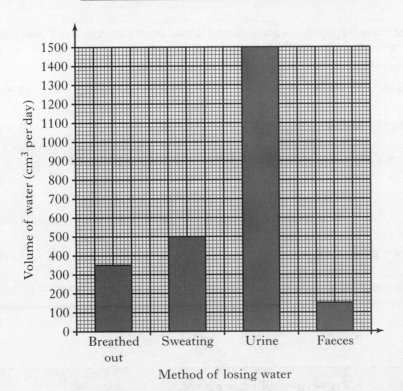

The graph shows Volume of water (cm³ per day) on the Y axis (0 to 1500) against Method of losing water on the X axis (Breathed out, Sweating, Urine, Faeces).

(b) Y axis labelled – volume of water in cm³ per day **1 mark**

Each bar drawn accurately **2 marks**

(c) Urea **1 mark**

(d) The bladder **1 mark**

TOP EXAM TIP

Make sure you draw the tops of the bars on the graph. You will not be awarded the marks if the bar is shaded and the horizontal top is not obvious. It is more important to draw these than to colour/ shade the bars.

8.

 This question covers Animal Survival (the need for food). It covers the digestive system.

(a) Starch and protein molecules are too big **1 mark**

and insoluble **1 mark**

HINT So need to be digested to pass into the bloodstream.

TOP EXAM TIP

Look at how many marks a question is worth. If it is worth 2 marks then make sure you make at least 2 separate points.

(b) (i) E
(ii) F **2 marks**

9.

 This question covers the Investigating Cells topic (Enzymes). It involves reading a graph and some KU about the characteristics of enzymes

(a) 50 mg **1 mark**

> **HINT** The first line is the pepsin one and you need to read along the x-axis to pH3 and then go up to the line and see what the reading is on the y-axis.

(b) (i) Pepsin – 2.5 **1 mark**

(ii) Amylase – 7 and 8 **1 mark**

> **HINT** There are 2 answers for amylase and both are necessary to get the mark

(c) Protein **1 mark**

(d) Less substance produced **1 mark**

> **HINT** At a high temperature the enzymes are denatured so less substance would be produced.

10.

 This question is a Problem solving question

(a) 3.1 kg **1 mark**

> **HINT** If she smokes 13 cigarettes a day then the bar is 11–15, the reading on the y-axis is 3.1.

(b) The more cigarettes you smoke, the lower the average birth weight. If none smoked birth weight average is 3.5 kg and if 21–25 are smoked the average is 3.06 kg. **2 marks**

(c) 0.35 kg **1 mark**

> **HINT** 3.5 – 3.15 = 0.35

(d) Bar completed correctly- **1 mark**

TOP EXAM TIP

When asked to interpret patterns in a graph try to use values in your answer as in (b), especially if worth more than 1 mark.

11.

 This question covers Body in Action topic (Movement). It covers parts of a joint.

(a) (i) A **1 mark**

> **HINT** This muscle must contract to lift up the lower leg.

(ii) D **1 mark**

> **HINT** This muscle contracting will pull the foot down

(b) Tendons **1 mark**

(c) Hinge **1 mark**

> **HINT** The function of ligaments, cartilage, tendons and muscles should be known.

12.

 This question covers the Body in Action (Need for Energy). It has both KU and PS questions.

(a) day 3 **1 mark**

> **HINT** This is when the reading was at its lowest.

(b) 1. Healthy person's reading is higher than asthmatic

2. Healthy person's reading is fairly constant and asthmatic's goes up and down more **2 marks**

(c) Yes

Reason: a reading of 230 matches the person with asthma. **1 mark**

> **TOP EXAM TIP**
>
> It is not sufficient to say that the healthy person's readings are high – you must have some kind of comparison – i.e., it would be enough to say that healthy person's readings were *higher* as this indicates a comparison,

(d) (i) He needs more oxygen/ needs to get rid of carbon dioxide **1 mark**
(ii) Lactic acid **1 mark**
(iii) Recovery time **1 mark**
(iv) short **1 mark**

13.

 This question covers the Body in Action topic (Co-ordination). It is a KU question

(a) A – Focus light rays

B – React to light

C – Carry nerve impulses **2 marks**

> **HINT** A is the lens   B is the retina   C is the optic nerve
> You should also know the cornea and iris.

(b) Easier to judge distances   **1 mark**

(c) Spinal cord   **1 mark**

 HINT   The 3 parts of the nervous system are brain, spinal cord and nerves

14.

   This question covers the Inheritance topic (Variation)

(a) 0.7 – 1

1.2 – 9

1.3 – 11   **2 marks**

(b) Y - axis labelled Number of berries   **1 mark**
Missing bars completed   **2 marks**

(c) 1:2   **1 mark**

HINT   The two values are 2 and 4, so these simplified make a ratio of 1:2.

15.

This question covers the Inheritance topic

(a) A group of organisms who can interbreed to produce fertile offspring
                                                                  1 mark

HINT   Organisms in the same species look similar to each other **and** can interbreed to produce fertile offspring. Both pieces of information are needed

(b) Continuous   **1 mark**

(c) Selective breeding   **1 mark**

16.

(a) The XY   **1 mark**

HINT   Males have XY and females have XX

(b) Sex cells would have one of each pair of genes   **1 mark**

(c) Down's syndrome   **1 mark**

(d) Amniocentesis   **1 mark**

17.

> This question covers Biotechnology (Living Factories)

(a) Glucose   **1 mark**

> *HINT* Fermentation equation   glucose ⟶ alcohol + carbon dioxide

(b) (i)   103   **1 mark**

> *HINT* 99 + 108 + 102 = 309
>   309/3 = 103.

> **TOP EXAM TIP**
> Repeating an experiment always makes it more reliable

(ii) more reliable   **1 mark**

(iii) As temperature increases the rate of fermentation
      increases   **1 mark**

18.

(a) 1. temperature / time / volume sugar / volume yeast/

         Any 2   **2 marks**

(b) A and C   **1 mark**

     Reason   – both of these produce a gas   **1 mark**

19.

> This question is a Problem Solving question.

(a) (i)   Amount of rice increases   **1 mark**
     (ii) Weed cover decreases   **1 mark**

(b) Competition for nutrients   **1 mark**

20.

> This question is a Problem Solving question.

(a) Fever, sore throats, coughs **1 mark**

(b) 76% **1 mark**

 HINT   42/55 x 100 = 76

(c) The flu vaccine is given to 'at risk' people **1 mark**

(d) A chemical which prevents the growth of micro-organisms **1 mark**

21.

★ This question covers Living Factories in the Biotechnology topic.

(a)

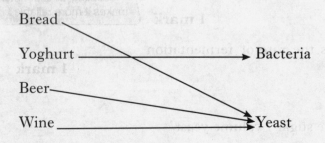

PRODUCTS   MICRO- ORGANISM

Bread

Yoghurt ————————→ Bacteria

Beer

Wine ————————→ Yeast

2 marks

(b)

STATEMENT	TRUE	FALSE	CORRECTION
Insulin is used to prevent the growth of micro organisms		✓	Antibiotics
Biological detergents contain enzymes	✓		
Lactic acid is important in the production of bread.		✓	Yoghurt

3 marks